Time
and the Physical World

RICHARD SCHLEGEL

Professor of Physics

Michigan State University

Dover Publications, Inc., New York

Published in Canada by General Publishing Company, Ltd., 30 Lesmill Road, Don Mills, Toronto, Ontario.
Published in the United Kingdom by Constable and Company, Ltd., 10 Orange Street, London WC 2.

This Dover edition, first published in 1968, is an unabridged and corrected republication of the work first published in 1961. This reprint contains a new preface by the author and is published by special arrangement with the original publisher, Michigan State University Press.

Library of Congress Catalog Card Number: 68-55820

Manufactured in the United States of America
Dover Publications, Inc.
180 Varick Street
New York, N. Y. 10014

To my Mother

Preface to the Dover Edition

I have made changes in the text in a few places for this edition, where new developments seemed to demand them, and also I have in a number of instances added or altered references, in reflection of new material that has appeared in the literature. But in the seven years that have passed since the original edition was published there has been—considering how rapidly we usually judge events to move in science—surprisingly little additional knowledge with respect to basic questions about physical time. I can hope, therefore, that the book retains what merit it may have had as an account of our time concept in the physical world.

Professor Robert B. Lindsay of Brown University has given me several critical suggestions with respect to textual alterations, and I am pleased to acknowledge his kind assistance.

R. S.

East Lansing, Michigan, 1968

Preface to the First Edition

This book presents a discussion of the time concept as it is used in our understanding of the physical world. That is, the concern of the book is with time as an aspect of the natural universe. Our approach is that of philosophical empiricism, and we do not assume an *a priori* time sequence with which natural events are to be associated. Instead, we study those events in order to discern what properties they have that make a time sequence so pervading and powerful an aspect of the natural world. Since our primary assumption is that the characteristics of physical time are determined by the processes of the physical world, we look to physical theory and observation for information about such problems as the uniformity, extent, and reversibility of time.

We shall not discuss in any detail the questions which relate to psychological time. These questions are concerned chiefly with the subjective quality of time sense, and with the quantitative and qualitative variations among different persons in their time perceptions. The psychological questions may well be of greater interest than the questions relating to physical time, but physical time is prior to psychological time, in the sense that there are physical processes in the absence of psychological processes, but no psychological processes without some physical process. In a sense, of course, our knowledge of physical time is psychological, in the way that all knowledge is psychological, but the point of difference is that our inquiries in this book will be directed toward the processes of the natural world generally rather than toward the psychological processes by which we respond to the temporal aspects of that world.

The first chapter of the book is devoted to the origin of our time concept in factors of change and regularity in our natural world. The relation between the direction of time and the thermodynamic concept of entropy is discussed in the second chapter, and an equivalence is found between the results of the analysis of time in the first chapter, based on general experience, and the analysis that utilizes the entropy concept. The temporal reversibility and irreversibility of natural processes are considered in detail in the next two chapters. In Chapter Five the analysis of the preceding four chapters is used in giving a discussion of the observed temporal characteristics of our universe.

Chapters Six and Seven are devoted to modifications of our concept of time that have resulted from the theory of relativity. The experimental confirmation of the relativistic time changes is discussed in detail, and the problems associated with extensions of these changes to clocks or biological organisms are considered. In Chapter Eight an hypothesis is suggested which completely maintains the physical content of special relativity theory but removes the problems of relativistic time changes for clocks or other systems that are not in situations of definite relativistic physical interaction. The effects of the quantum theory on our time concept are discussed in Chapter Nine. The final chapter is in the form of a dialogue; the participants, a physicist, a philosopher, and a physical chemist, summarize many of the arguments of the book, and also permit themselves some speculative discussion.

It is hoped that this book may be of interest to students, professional or amateur, of the philosophy of physics. An attempt has been made to present the concepts of statistical mechanics, relativity theory, and quantum theory that are used, so that a reader without prior knowledge of these subjects will still find the discussion meaningful. Much of the book may therefore be regarded by a trained physicist as popular or semi-popular exposition of science. On the other hand, the reader who has had no college training in physics will probably find the material of many parts of the book to be highly unfamiliar. In Chapter Three the elementary concepts and operations of the calculus are freely used, but elsewhere the mathematics, with a few exceptions, is generally confined to ordinary algebra. In any case, I believe it may be fairly said that the arguments and discussions of the book are primarily discursive rather than mathematical.

There are many teachers and writers to whom I am indebted, and I wish that I could thank each one individually. I do wish specifically to express my gratitude to Herbert Feigl and to William H. Sheldon; to the former for what he taught me of the role of philosophy in intellectual activities, and to the latter for his encouragement of my scientific education.

Most parts of the manuscript of this book were read by Professor J. H. McKee, of the Department of English of Purdue University, and I am indebted to him for many suggestions—and corrections—with respect to my grammar and diction. Professor Henry Margenau, of Yale University, has kindly read the manuscript; I am grateful to him for much valuable criticism, and I wish to thank him for his support at a critical time in the life of the manuscript. My colleague in the Department of Physics at Michigan State University, Professor D. J. Montgomery, has read the entire manuscript in an early draft, and I want to thank him for the many improvements he suggested, both on points of style and of content. I also wish gratefully to acknowledge the valuable and careful criticism which I received anonymously from a publisher's reader.

For major assistance in the work of preparing the manuscript, for discussions on almost every section of the book, and for her sympathetic encouragement, I am deeply grateful to Sally McKee Schlegel.

R. S.

East Lansing, Michigan

Table of Contents

Chapter I

Time and Physical Process

1. TIME AND PHYSICAL CHANGE.

Time is associated in an essential way with the changing characteristics of the natural world. If a region or village has seemed not to change we say that for it "time has stood still." This very usage reflects the idea that the changing of the world is a concomitant of the passage of time.

We might ask, Does the world change because time moves on, or does time move on because there is change? The answer must be that time arises from changes that occur. To answer otherwise would be to assume that there is a time entity that is prior to the events of the natural world. We are here assuming contrariwise, that time is an aspect of the natural world, co-existent with that world but not prior to it. Evidence against the existence of time that is prior to events will be given later in this chapter.

It is not difficult to perceive that the existence of time is related to the existence of change, and yet, it may seem that change is not the essential element of time. A quality of persistence or extension—of "everything not happening at once"—seems to be the most fundamental and essential characteristic of time. That is, if the natural world changes it presents a succession of different states or conditions. The capability of presenting such a succession requires that nature possess a property of taking on or displaying more than one state. This property might be called "capability of showing change," and we could then say that we find time in the world because the world possesses the "capability of showing change."

It should be made clear, however, that the *possibility* of presenting a succession of different states would not ensure the existence of physical time. We can think of a universe in which there is no change whatsoever, although there is a possibility of change being introduced; such a universe would be one which possesses the "capability of showing change" but in actuality presents only a single state. In this universe there would be no time. An intuitive feeling that a changeless universe would in fact still have a time progression associated with it is an expression of our mental habits. The world we live in indubitably does present a succession of different states, and we are therefore accustomed to considering time and change to be necessary attributes of existence. But a universe which in *all* of its parts showed no change whatsoever would not have a time measure associated with it, for there would be nothing about the universe that could distinguish one state as being at a different time than any other state.

We will say, then, that change in the natural world is essential for the existence of time. We realize that "capability of showing change" is a more fundamental natural property than the actual showing of change, but since such a capability is not alone sufficient for the existence of time, we will stipulate natural change as the first element in the constituting of time.

Some hypothetical examples of the physical situations which we have suggested may be helpful. A universe which presents states which are successively different is well known to all of us; for on levels of observation from the ultra-microscopic to the astronomical the natural world about us is one of changing properties. A large mass of homogeneous matter, idealized as being in mechanical and thermal equilibrium and with no interaction of any kind with the surrounding universe, would be a domain in which there is no change of any kind. Hence *within* that domain we could even say that there is no time, although the domain be surrounded by a changing world in which there is proper time measure. The introduction of, say, a heat source into the domain would give rise to a succession of different states as the material continually grew warmer; this succession would constitute an element of change, and therefore a time element, in the domain. The entire universe, similarly conceived as homogeneous matter in an equilibrium state, would be a universe possessing the "capability of change" but with no time property.

To illustrate the meaning of a universe without the property of presenting successive and different states we might think of a world which has only a momentary existence, in terms of the time sense of our actual universe, or of a universe which is such that changes cannot occur in it. A physical world which existed, for instance, only as do the photons in a flash of light which is quickly absorbed would be a world with little possibility for existing in different states. Or, a universe in which the constituent entities—the atoms say—could not possibly move relatively to each other and could not take up or lose energy of any kind would also be a world which did not possess capability for change, and hence would be a world in which time did not exist. Such a world, of quiet atoms in constant energy states, is of course far different from the actual world about us.

2. CLOCKS AND CYCLICAL PROCESSES.

Physical change of any kind whatsoever may be considered as giving rise to a time that is not yet differentiated, or not yet measured in defined units. For the setting up of a measure of time passage we use physical change of a special kind: we define time units in terms of processes of change which are *cyclical* in a uniform manner. A cyclical process of a physical system (by system we merely mean a chosen group of material entities) is a process in which the system repeatedly goes from some initial state 0, through intermediate states 1,2,3,........n, and then again to the original state 0. A selected cyclic process may be taken as a basis of time measure; the time unit is defined as the duration of one (or any other chosen number) of the cycles of the process, and the time extent of a non-cyclic process of interest may be determined by finding how many cycles of the cyclic process occur between the initial and final states of the non-cyclic process. The cyclic process that is so used as a time measure is the essential process of a *clock*.

The first cyclical process to serve man as a time measure was almost surely the rotation of the earth on its axis. The alternation of day and night presents a natural time measure for the non-cyclical changes of nature, and historically the day must have been the first time unit. The seasonal changes associated with the orbital revolutions of the earth, and the phase changes in the moon resulting from its revolution, have of course provided two other obvious natural cycles that have long served as units of time measure.

In our society today we use a variety of cyclical processes as bases for clocks. The pendulum clock uses a simple cycle of back-and-forth mechanical motion. The rate of an electric clock rests on the cyclical electrical potential variations in the supplied electric power, those variations themselves being a reflection of the revolutions of the moving elements in the power house alternators. Some clocks of high accuracy utilize the semi-elastic oscillations of a crystalline material as the basic cyclic process, and the most accurate clock at present available—the cesium [1] clock—depends for its rate regulation on the precessional motion of nucleus and outer electron in the cesium atom.

Whatever natural cycle is used for a clock, it must satisfy the primary condition of being a uniform cyclic process; that is, any one cycle must be the same as any other. If we have an established cyclical process which we have set up as a satisfactory standard time unit, we can say of any other cyclic process that it is uniform if

[1] L. Essen and J. V. L. Parry, *Nature, 176,* 280 (1955); Phil. Trans. Roy. Soc. A, *250,* 45-69 (1957). See also "Atomic Clocks," by Harold Lyons, *Scientific American,* February, 1957.

each of its cycles occurs in the same interval of time. The cyclic processes which are used in good clocks obviously must have this property of uniformity. A cyclical process like the revolving of a wheel on an axle, such that the wheel first spins at a high angular velocity but then because of frictional effects gradually rotates more slowly, would obviously not be a uniform cyclic process.

The establishing of a standard time unit based on a uniform cyclical process raises the question, How may we know when the chosen process is uniform? If there is no time standard prior to the standard process which defines the time unit, may not that standard process actually vary from cycle to cycle? The answer to this question is two-fold. In the first place, any process chosen to be the standard one for defining the time unit will be examined for any possibility of known physical influences which might cause one cycle of the process to be different from another. Thus, the axial rotation of the earth is known to be not an exactly uniform cyclic process because, for one reason, the ocean tides constitute a kind of earth to moon-and-sun frictional effect that over a long period of time affects the earth's axial rotation. Likewise, it has been found that the earth's orbital rotation about the sun is subject to slight perturbing influences. A clock based on an atomic cyclical process, on the other hand, is presumably based on a process which can be maintained virtually free of perturbing interference. On physical grounds a crystal or atomic clock, therefore, is in principle to be preferred as an ultimate time standard over an astronomical cyclic process; the atomic process is a simple or "pure" natural process, whereas a process like the earth's rotation is compounded of one major cyclic process with several minor processes that become not quite trivial when a very high degree of uniformity from cycle to cycle is desired. In practice, however, we are by no means ready to give up astronomical cycles as standards for time measurement.

But, the consideration of directly physical criteria is associated with a second way of determining whether or not a given cyclic process is uniformly cyclical. Since there are cyclic processes in nature, we can test the uniformity of any one process by observing the consistency of its cyclic frequency with the frequency of other presumably uniform cycles. Thus, we find that the axial and orbital rotation frequencies of the earth, the frequency of revolution of the moon, the frequency of a constant amplitude pendulum swing, and many atomic and electromagnetic wave frequencies all bear to a certain degree of accuracy a constant ratio to each other. At any investigation that we make we find, say, that a given pendulum swings through 86,400 cycles while the earth revolves once on its axis. Or a kind of yellow light is consistently found to result from 4.4×10^{19}

electromagnetic wave oscillations during a suitably determined revolution of the earth on its axis.

These various cyclic phenomena seem manifestly to be uniform in their cycles because the duration of a cycle of any one of the phenomena is some constant fraction or multiple of the duration of a cycle of any other of the phenomena. Having established a group of cyclic phenomena—that is, that group of processes whose cycles are in constant frequency or time ratio to each other—we can determine whether or not any newly considered phenomenon is a cyclic process by observing whether or not its cyclic frequency is of constant ratio to any of the processes of the group. Thus the simple timing with a clock of a rotating wheel establishes whether or not the wheel's rotation is a uniform cyclic process; the process of the clock which is being used has presumably already been established as a member of the group of natural uniform cyclical processes.

What we are saying here is that by a consistency criterion we distinguish those processes which are uniformly cyclical from natural processes generally. The physical criteria for uniformity may then be applied to members of the cyclical group, so as to determine which are most free from disturbing physical influences and will therefore present in their behavior virtually only a single and uniform cyclical natural process. In practical work, the consistency condition is determined not so much by direct observation as through the mediation of a physical theory: the cyclical motions that are accurately described by theory are members of a consistently uniform set of motions (assuming, of course, that the theory is consistent). This point has been well discussed in connection with man's actual time-keeping by G. M. Clemence, who states:[2] An invariable measure of time is . . . "a measure that leads to no contradiction between observations of celestial bodies and the rigorous theories of their motion; yet more precisely, it is the measure of time defined by the accepted laws of motion."

Clemence points out that the irregular fluctuations in the earth's rate of rotation have been established by comparison with other planetary or satellite motions. "Among the motions in the solar system that can be observed accurately enough to be used as clocks are the rotation of the earth, the earth's revolution around the sun, the revolution of the moon around the earth, that of Mercury and Venus around the sun, and the four large moons of Jupiter around that planet. Of these nine clocks it is found that eight agree with one another, and the outstanding disagreement is the rotation of the earth."[2] The decision that it is the earth's axial rotation which is irregular rests on the small likelihood that every one of the other

[2] "Time and Its Measurement," *American Scientist, 40,* 260 (1952).

eight motions would be irregular by the same amount of time (the amount of the earth's irregularity). One can of course then look for physical reasons for the earth's irregularity, and in finding such one further strengthens the choice of the other motions as the properly uniform ones. But in any case—and this is the relevant point—the uniform cycles are chosen as defining a time unit because with that unit the entire theory of celestial motions becomes consistent with observation (or, if one likes, the set of theoretically expected uniform cycles are then found consistently to be so).

We can argue that our choice of time unit is arbitrary, and we have merely *defined* which cycles constitute the set of uniform cycles. In a non-practical sense this is true: any cycle, even one that is highly irregular by our present time standards, could in principle be chosen as setting our time standard. Various other cycles would then be found to vary in odd ways; each day, for example, might differ widely from its predecessor in length. But for the actual description and theory of celestial motions, time units have been used which give us a large group of cyclical motions that are uniform from one cycle of a motion to the next, as compared with other similarly uniform motions and in accordance with physical theory. The uniformity that is found rests in part, then, on theory, which may be considered as a kind of extended definition, but also in part on the nature of our physical world. Conceivably, the motions of all physical bodies might be so chaotic that no theory could be found, and no associated time unit, which would enable us to find cyclical processes that possess any kind of uniformity. In fact, as has been noted, it is otherwise: with a theory of motion that is correctly descriptive, at least as far as it goes, we do find astronomical and atomic cycles which are among themselves consistently uniform from one cycle to the next.

It may be objected that the consistency criterion for cyclic uniformity is invalid, in that *all* of the cycles of the group of uniform cycles might be varying in tempo. Thus it has been proposed that time might everywhere "dilate" or "contract," and yet we would be unaware of the change because of the uniformity of the effects; if all processes began going only half as fast as formerly, our mental processes as well as all physical processes would be equally affected and the world would seem quite as before. Such a hypothetical *universal* variation in time rate is actually of uncertain meaning, for a proposed time change which is in principle not observable would be without significance except in a context of physical theory. We shall, however, consider this possibility in two specific ways, and shall find that in neither case is our criterion for uniformity invalidated.

First, we might think that every time-keeping process is speeded up, so that, for example, a given revolution of the moon about the earth which previously required T seconds now takes only $T/2$ seconds. All other processes would be similarly quickened. This speeding-up may be regarded, however, as nothing more than a re-definition of time units. Since our time unit has only the warrant that it gives us consistency or uniformity among cycles as prescribed by theory, there is no physical import to the question of what is *the* length of a time unit; for there is no process over and beyond the natural systems of our world in terms of which that length may be defined. We might re-define our time unit, choosing the "dubsec" in place of the second, with one "dubsec" equal to 2 seconds. Time values, expressed in the new unit, would numerically equal one-half of time values in terms of seconds, but there would be no disturbance of the consistency relations that had been found, in terms of the chosen time standards, among natural cycles.

Second, we might in a speculative way consider that some time-keeper exists outside of the domain of the natural universe, and that by the measure of his time processes the tempo of all of the physical processes of nature is changed. We now find that such an universal variation in our chosen group of uniform cyclic processes would require drastic revisions in natural law. As an example, let us assume that all physical processes are to be doubled in tempo, so as to occur in "half of their previous time." A swinging pendulum, then, for which the frequency is $(\frac{1}{2}\pi)\sqrt{g/l}$, where g is the gravity acceleration of a mass and l is the length of the pendulum, would begin to swing at twice its previous rate provided that the gravitational constant g were quadrupled in value. But also the velocity which a body reaches in time t in falling from rest under the influence of gravity is gt; that is, the velocity is proportional to g and not to $\sqrt{g}$. Therefore, the fourfold increase in g which caused the frequency of the pendulum swing to be doubled would cause the velocity acquired by a falling body to be quadrupled, in a given time t as measured by the "outside" time-keeper. Within our natural domain, to be sure, a change in the value of g would not bring temporal inconsistencies, but for the hypothetical outside observer the change would, we see, have varying influences on the time-rates of different processes.

Countless other illustrations may be devised to show that we could not get a universal change in all physical processes and maintain our present form of physical law. We could not do this because different processes do not in general depend on the same mathematical power of the various physical constants. Hence the change in constants which would be required to give changes in tempo would of necessity give different changes to different processes. A somewhat more involved example of dependence of time rate on physical constant than the pendulum-falling body illustration is provided by the effects of the value for the heat energy released by the oxidation of carbon. This heat energy is perhaps a factor in the time rate of human physiological performance, and it is certainly a major factor in determining the rate at which an automobile moves. But the increase in oxidation energy value which would double the speed of a car motor (and which would have to vary from car to car!) would certainly not have a doubling effect on the speed of physiological processes. We may indeed conclude that the notion of a physical but undetectable universal change in time-rate of all natural processes has the characteristics of an absurdity.

3. NATURAL LAW AND CYCLIC PROCESSES.

The uniform cyclic processes of nature provide us, then, with a clear and definite time measure. Non-cyclic processes of change may be given a time length, by reference to the cyclic processes in terms of which a time unit is defined.

Time's apparent quality of being all-pervading and of possessing an essential evenness of flow rests essentially on the existence of uniform cyclic processes everywhere in nature. Were there not such cycles presented to us, the passage of time would be associated with non-cyclic change, and time would certainly have a haphazard, irregular quality—resulting from things happening now rapidly, now slowly, and differently at one place than another.

The fact that there are cyclic processes which give rise to our time concept must be considered to be one of the major generalizations that we can make about the natural world. The fact that the various uniform cyclic processes do bear a constant frequency ratio to each other is a further general property of nature of the widest significance. This fact is related to what is sometimes expressed as the "uniformity of nature" or the "order of natural law." What we do find is that two given cyclic processes, say the orbital revolution of the earth and the oscillating motion of a prescribed pendulum, have within a certain narrow range the same frequency ratio at any investigation that is made of the ratio. A similar constancy of ratio is found for other cyclic processes, in atomic as well as terrestrial and astronomical domains. The constancy is so universally found for a given ratio of cyclic frequencies that we confidently assume the constancy to exist everywhere in the universe. Further, we confidently and usually justifiably assume that non-cyclic processes have a similar uniformity, such that similar natural processes will always occur in the same time interval, as measured by standard cyclic processes.

We see, then, that our experience of time as regular, as having equivalent units, rests on the uniformity of repeated process that does exist in nature; or, we can say that our concept of time would not be as it is were it not for the order of nature. The orderliness of nature may be said in turn to rest on the fact that natural processes occur in accordance with descriptive natural laws which mankind has been able to discern. These laws take various forms, and the set of natural laws is certainly not unique; there are alternative sets of laws by which the phenomena of a given domain may be described and predicted with a high degree of accuracy. Yet, for any given domain of experience, there is usually a preferred set of descriptive natural laws, and these laws—as accurate descriptions of nature—describe the cyclic and non-cyclic phenomena that are observed to occur.

We have discussed in the previous section the possibility of a universal, consistent change in the frequencies of all cycles, and we saw that the possibility of such a change is physically unreasonable. A different possibility, however, is that one or some of the cyclic processes of nature might undergo change in frequency, as a result of slow changes in the laws of nature or in the fundamental physical constants. Such changes would of course have discernible effects, in that there would be resulting changes in the frequency ratios of various processes. It is a matter of speculation whether or not such changes occur; certainly the customary assumption of physicists in their work is that the laws of nature are not changing.

There are, however, cosmological theories in contemporary physics which do require that as the universe evolves, in the manner presented by the theory, there will be relative changes in the values of the physical constants.[3] For example, the late E. A. Milne devised a cosmology which required in effect two time scales, one for electromagnetic processes and one for mechanical processes, to state briefly the distinction between the two scales. At various stages in the development of the universe the ratio between the scales would differ, with the result that, in effect, the frequency ratio of a given kind of light waves and a prescribed mechanical oscillator would vary. In quite a different manner Professor P. A. M. Dirac has presented a theory which relates the gravitational constant to other basic constants of the universe, in such a way that with the presumed expansion of the universe the gravitational constant would be decreasing in magnitude. A pendulum clock, as timed by an atomic-oscillations clock, would therefore have run faster a billion years ago than today; or, a solar system of the same masses and kinetic energies as our present system, would have had different orbital sizes and frequencies a billion years ago than it has to-day.

A further development of the Dirac suggestions has been carried

[3]See H. Bondi, *Cosmology*, Cambridge University Press, 1952. Also, E. Teller, "On the Change of Physical Constants," *Physical Rev.*, *73*, 801 (1948); R. H. Dicke, "Principle of Equivalence and the Weak Interactions," *Revs. Modern Phys.*, *29*, 355 (1957), and "Gravitation—An Enigma," *American Scientist*, *47*, 25 (1959).

out by Pascual Jordan. He assumes a total constant energy for the universe, with a balance between the (negative) gravitational potential energy of a body and its positive mass-energy; new matter may then enter the universe, without violation of conservation of energy. In Jordan's theory, the universe is steadily growing in total mass and in size, and, as in the Dirac theory, gravitational attraction between particles is growing smaller relative to electrical attraction.

4. ABSOLUTE TIME.

The possibility of changes in physical laws or constants—although important for the problem of the age of the universe—need not detain us longer at this point. From any practical considerations of time-keeping, even of the highest accuracy, the possibility of changes in frequency ratio is unimportant, since no discernible changes are expected in any theory except over periods of millions of years. The theoretical possibility does emphasize, however, the manner in which we do base our time concepts on the cyclic processes which exist in nature. We can speak of the frequency of some one cyclic process changing with respect to the frequency of another process, but we cannot speak in any absolute sense of a cycle's changing in frequency, because we have no absolute measure of time.

Until a few decades ago the view was common among physicists, though not without its critics, that there existed in nature a "stream of time" which moved evenly forward and provided a common coordinate to which all events could be referred with respect to their time relationships. This view seems to have had some acceptance in Greek times (as indicated by the *Timaeus* of Plato) and it was made a part of physical theory in the seventeenth century by the chief author of physics, Isaac N.B. Newton. Newton spoke of time as a stream that "of itself and from its own nature flows equably without regard to anything external."[4] Time so conceived is our *absolute time* which would exist in a world without change or matter. There is an implication in the notion of absolute time that there exists in some manner a great master clock, over and below the natural events of the world, and that this clock keeps a time that is everywhere flowing, independently of the events of the world.

We will see in a later chapter that the theory of relativity specifically invalidates the absolute time concept; for that theory predicts differences in rate of physical process, for a given physical system, under varying conditions of motion, and the existence of these differences has been confirmed experimentally. But even without referring to the theory of relativity we find reasons for rejecting the

[4]Isaac Newton, *Principia*, Scholium to the Definitions.

absolute time concept. We find cyclical processes in nature, but we find them to be of varying frequencies, and any one process may be varied in value, in practice or in theory, by variation of the physical parameters that characterize the system which undergoes the process. We do not, then, find any "master cyclic process" which serves as a master clock that is reflecting an all-pervading absolute time flow.

We might say that since there are uniform frequency ratios among the various cyclical processes—neglecting relativity theory effects—we may assume an absolute time which is regulating these processes. We, however, gain nothing by way of scientific explanation in making such an assumption, since there is no observable physical process, corresponding to that assumed absolute time flow, which is additional to the observed processes. On the other hand, in taking the point of view that time is an aspect of physical process, we can understand time as a part of the total unity—the natural world—which we are trying to understand. Further, certain hypothetical extreme conditions of nature—as a state of no change whatsoever—then carry with them changes in our time concept, as will be discussed in the final section of this chapter. On an assumption of absolute time—which is in fact not justified in any event by observation—all possible conditions and processes of nature must be fitted into a concept of time which cannot be altered, whatever the ultimate state of change or not-change in the universe may be found to be. To forbid any modification of a concept by further experience is altogether counter to proper scientific procedure.

N.B.

5. TIME MEASUREMENT AND NON-CYCLIC PROCESSES.

We have emphasized the role of cyclic processes in providing both clocks and a standard unit for time measurement. The question may well be asked, Could we not equally well use a non-cyclical natural process for a time measure? We can think of a clock's being made, for instance, from the natural decay of a radioactive isotope.[5] Such a process is uniform, in that a constant fraction of a given sample will always decay in a given unit of time. Or, somewhat more crudely, we might think of a process like the weathering of a rock as providing a time measure. A fairly constant stream of water, for example, might in each year wash away the same depth of rock over which the water is flowing.

In the examples cited, however, we think of a non-cyclic process as serving as a clock, on the basis of a calibration which utilizes

[5]J. A. Carroll has discussed in detail how a clock could be made of two radioactive materials with different decay constants: *Nature, 184,* 260 (1959).

a time standard derived from cyclical processes. The question which we wish to answer is whether or not non-cyclic processes alone could serve as primary standards and clocks in our time measurement. If this were to be done we would expect to be able to find in nature a number of processes which are non-cyclic but yet are undergoing change in some uniform manner; these processes would serve as clocks and time standards, as do in fact the various natural cyclical processes. And just as non-cyclic processes are satisfactorily described by time units and natural laws which are the same as those for cyclic processes, we would expect that if non-cyclic processes were the primary ones for time measurement the cyclic processes could be accurately described in terms of time units based on the non-cyclic processes.

Experience indicates, however, that uniform non-cyclic processes of change are not abundant in our natural world. Even approximately uniform non-cyclic processes, like the weathering of rocks, are not obvious, and non-cyclic processes, which depend on a few simple physical parameters and are therefore highly uniform, are rare in nature. The decay of radioactive materials is one such process, and the apparent recessional motion of galaxies, with a velocity directly proportional to their distance from the earth, is a second. Both of these processes are fairly subtle ones to discern, and it is safe to say that if man had had to depend on them as bases for clocks he would have been a long time indeed in conceiving of time as being in any way uniform and capable of quantitative division. However, the theoretical possibility of basing our time measure on one of the few uniform, natural non-cyclic processes cannot be denied. It is true that there are contrived non-cyclic processes which serve as clocks—the hour-glass and water clock are examples—but they were invented by men who had gained their time concept from the natural cyclic processes.

With respect to the concept of time, a striking feature of our universe is that it does present cyclic processes in abundance; and hence, the universe does readily suggest a time concept which implies persistence of entities through uniform, successive time intervals. Non-cyclic processes are ones in which a system is continuously being carried into states which are different from all previous states of the system. If all the processes of nature were of this kind—like the radioactive decay and galactic recession processes—the world would always be changing, and it would be perhaps difficult even to identify any persisting aspect of nature. The fact, however, that many processes are cyclic reassures us that absence of change is characteristic of at least parts of our universe, for in a cyclic process each state of a system is a return to a previous state. Those same cyclic processes then which give us a time measure are also essential for our changing universe's possessing elements of permanence and stability.

6. THE ORIGINS OF TIME.

We have seen that there are two essential physical requirements for our time concept: processes of change which give rise to successive differences in states of the world, and cyclic phenomena, among those processes, which give a basis for time measurement. If there were change only, without the cyclic phenomena, we would not have a time concept similar to the one we do have, for we would have no sense of time as being composed of uniform and equivalent units. We have also the theoretical, rather unlikely, possibility of a time measurement being based primarily on uniform, non-cyclic processes; with that possibility there are still two requirements for our time concept, but a group of uniform non-cyclic processes is substituted among the various processes of change for the group of uniform cyclic processes.

It is perhaps an aid in understanding the nature of our time concept to consider hypothetical extreme conditions in our world, with respect to the two requirements. If change alone, without the presence of a group of clearly uniform processes, will not give us our time concept, we may next ask, What would be the effect of there being uniform processes of change only? Our actual world contains uniform cyclic processes among the various processes of change. If the world contained only cyclical changes, would we still have our time concept? The answer must be, "No." For unless the various cyclic processes interacted or continuously changed their phase relations so as effectively to produce non-cyclic change, a world which was dominated by cyclic processes would continuously return to previous states and there would be no sense of a progressively increasing sequence. For illustration we might think of our terrestrial world as being such that there were no indications of geological evolution, so that the only observed changes in landscape were the seasonal ones, identically repeated each year. Further, we think of there being no biological evolution and no significant social change, so that each generation of animals and men remains as in preceding generations. There would then be a sense of time, and of going through a succession of different states, within each generation; but in the extreme case where each generation lived exactly as any other in a world that was the same for each generation, there would be no time that progressively increased from generation to generation.

In considering a purely cyclic universe, we must be on guard, as also in the instance of a universe of no change whatsoever, against the notions of time which have become intuitive with us. We tend to think that time always is going on, no matter what is happening in the universe. In fact, however, if the world went through equivalent cycles repetitively, there would be nothing to give rise to a concept of one cycle's being earlier or later than another, and there would be no criterion for distinguishing any one cycle from another. The changes within a cycle could give a sense of temporal succession, within the cycle, but beyond one cycle a purely cyclic process itself may be considered to be an atemporal process.

Any physical process that is purely cyclical may be treated as being an atemporal process as long as it is considered only in its own domain, without reference to the larger, temporal universe surrounding it. The successive oscillations of a photon as it moves through a vacuum, for example, are identical, and there is no change that gives a time coordinate. The motion of an electron about a nucleus, or the cycles of two isolated astronomical bodies rotating about each other, have a similar atemporal quality. On interaction with other systems these purely cyclic processes may become non-cyclic and thereby gain a temporal character.

The cyclic process, then, can become a temporal process only by reference to non-cyclic processes. We may refer to the latter as *progressively* changing processes, in contrast to the cyclic processes in which we do not have progressive change except within any one cycle. We have seen that the progressively changing process does not give us our time concept because change in general does not provide a basis for time standardization and measurement. But likewise, uniform cycles of change will not by themselves give us our time concept, because without progressive change there is no distinction of any one cycle from another. Time has its origin *in the existence of both kinds of physical change, cyclic and non-cyclic, in the natural world that we know.*

Conceivably, the world might be such that it contained progressive change, with a sufficient number of uniform non-cyclic processes that time measurement could be based on these uniform processes rather than on cyclic processes. Our time-concept, however, is so clearly based on uniform cyclic processes, in a universe that presents an abundance of such processes, that we shall neglect the hypothetical possibility of standardizing uniform non-cyclic processes and henceforth confine discussion to the cyclic processes in regard to the providing of time standards and clocks.

Our natural world does seem to be one which is obviously undergoing progressive change. The evolution of species, the geologic changes in the earth, the evidences of development in the stars: these are all processes of non-cyclic, progressive change. Even the life of any one human being is emphatically this way, with its differing stages of growth, maturity, and deterioration. And yet, throughout the natural fabric of progressive change there are woven countless cyclic processes, constant in their frequency ratios, in terms of which man can measure the stages of progressive change. We could state the matter with alternate emphasis, and point out that we find cyclic processes everywhere in nature, but also, along with these phenomena there are progressively changing processes which give us that sense of the world's moving from state to state—or from time to time—by which we are able to discern two successive cycles as occurring at different times.

The problem of the roles of progressive and cyclic change in the physical world will occupy us in later chapters; we shall consider, in the light of physical theory, possibilities for the world's being fundamentally progressive and hence temporal, or basically cyclical and hence not "in time" in a thorough, overall sense. Here, however, we merely observe that the two elements of time enter into our ordinary experience, so that to some extent we are acquainted with both a temporal and an atemporal world.

Progressive change is perhaps the dominant element in our experience, of the two that give rise to our time concept. And yet to a small child, for example, whose life is evenly passing from day to day, there is almost a complete absence of a sense of time's "moving on." Rather, the world is made of units each very much like the other, and later experience must teach the awareness of progression and change that comes with adulthood. Even in one's adult life, and in a society that changes as much from one generation to the next as Western societies have done in recent decades, contemplation of the repetition of generations brings some sense of time's not moving on, of the atemporal quality of cyclic process. For one's parents and grandparents, and so on back, in many essential ways lived much as we do, facing the same situations of life that are the lot of us today.

Contrariwise, we well know that our grandparents' generation is earlier in time than ours of today, because of the progressive changes that have occurred in the intervening years. Likewise, we confidently assign a later date to each successive revolution of the earth about the sun; each of the cycles of that process, in our scientific experience, is indeed closely equivalent to any other, but progressive changes from cycle to cycle on the earth, as well as in the universe at large, prevent the earth's revolving from being a purely atemporal process. And so it is throughout our experience: the natural world presents both progressively-changing and cyclic processes, and we have a time concept which implies both a going-on change of the world and a presence of equivalent time units for the measuring of any process of change.

It has sometimes been argued that the origin of time rests on the existence of a causal relationship among all physical events. This point of view has been stressed by Hans Reichenbach, in his book *The Direction of Time,* in which he states that ". . . time order is reducible to causal order." [6] In one sense a physicist can scarcely disagree with such a statement, because we would not accept a time order among events if the order violated physical causality relations. But on the other hand, it clearly is not necessary (and I do not believe Reichenbach wished to imply this) for us to know what the causal connections are between events if a time concept is to be established by those events: primitive man achieved a notion of time, without any very exact knowledge of causal relations between events.

In this chapter we have discussed the foundations in physical phenomena of our time concept. These are the bases, we believe, on which the concept does in fact rest, and it is because of these characteristics of the physical world that we have our time concept. Further, it seems evident that this concept is prior to any detailed knowledge of physics.

General References for Chapter I.

M. F. Cleugh, *Time and Its Importance in Modern Thought,* Methuen, London, 1937; Louise Robinson Heath, *The Concept of Time,* Univ. of Chicago Press, 1936. These two books give presentations of various philosophical viewpoints with respect to time.

G. M. Clemence, "Astronomical Time," *Rev. Modern Physics, 29,* 2 (1957). The merits of primary astronomical time standards as compared with atomic time standards are discussed by Clemence, *Nature, 176* 1230 (1955), by A. Perard, *ibid., 177,* 850 (1956), and by L. Essen, *ibid., 178* 34 (1956).

A. C. Benjamin, *An Introduction to the Philosophy of Science,* Macmillan, New York, 1937, Chap. XIV.

J. T. Fraser, ed., *The Voices of Time,* George Braziller, New York, 1966.

Adolf Grünbaum, *Philosophical Problems of Space and Time,* Alfred A. Knopf, New York, 1963.

Henry Margenau, *The Nature of Physical Reality,* McGraw-Hill, New York, 1950, Chap. 7.

Hans Reichenbach, *The Philosophy of Space and Time,* trans. by Maria Reichenbach and John Freund, Dover, New York, 1957, Chap. II.

Bertrand Russell, *Human Knowledge: Its Scope and Limits,* Simon and Schuster, New York, 1948, Part Three, Chap. V; Part Four, Chap. V.

A. N. Whitehead, *The Concept of Nature,* Cambridge Univ. Press, 1920, Chap. III.

G. J. Whitrow, *The Natural Philosophy of Time,* Nelson, Edinburgh, 1961.

[6] H. Reichenbach, *The Direction of Time,* the University of California Press, 1956, p. 25.

Chapter II

The Direction of Time

1. PAST AND FUTURE.

One of the major characteristics of our concept of time is that time is considered to "move forward," in a consistent sense, from past to future. Analogically we may say that we think of time's being a long road on which the world moves, with the already traversed past stretching behind and the future running out ahead as far as can be seen. The movement of the traveller-world is always without reversal of direction, and the road does not come back upon itself.

It might be argued that a statement about time's direction is without significant content. The changing world goes through a succession of states, and the direction of time, we might say, is merely as given by the passage from one state to the next; the succession of states, whatever it may be, determines the direction of time. It is tautological then to speak of time's direction as being "forward," or from past to future, for any succession of states that does occur will by definition be from past to future. We place any states that have occurred in the past, and the ones that are yet to occur constitute the future.

The statement that time moves from past to future is, however, more than a definition of those terms, and is in fact a statement about the nature of our world. The physical content of assertions about time's direction is apparent when we consider possibilities for temporal behavior different from that which the world actually does show. We might suppose, for instance, that in our experience the world did repeat itself in great cycles. The cycles would presumably involve human behavior and memory, but by scientific inference the cyclical aspect of the world might be discerned. Quite possibly we would then choose some state as a beginning point in the cycle, and at this point say that time moved into the past—that is into a cycle previously traversed. Or, we can think of a world which occasionally immediately after having been in states 1, 2, 3, ...*n*, retraced its steps and took up states *n*,........3, 2, 1, and then again states 1, 2, 3,........n before going on to state $n + 1$. These are fantastic speculations, in view of the kind of behavior that our world does show, but they indicate kinds of behavior which would lead to our characterizing time as other than always moving forward.

We commonly find, then, that time has a unique past-to-future dimension because our world goes through successive states each of which differs from its predecessors. One of the great generalizations of physics is that a physically defined quantity, the entropy of an isolated system, increases with time; that is, increases as the system goes from one state to the next. We shall discuss this generalization later in this chapter, but first we shall discuss with some care the association between time and successive states. It is desirable to do this if we wish to present entropy considerations as giving insight into characteristics of time that have already been established on the basis of common experience. We do not want to make the mistake of writing as if mankind had not known about the passage from past to future until physicists introduced the entropy law.

We consider a group of n states of the world. The general problem of time direction or time order is: what are the characteristics of states whereby they may be ordered in a series running from past to future? An individual person orders the states of the world of his experience in terms of his own time sense, and we might rest the determination of a time order on the existence of such personal earlier and later concepts. We would however not then have a criterion for time's direction in terms of states of the physical universe, and such a criterion is desirable as a corrective against individual peculiarities and aberrations in time sense.

We associate a time number θ with each state of the world, using the cyclic clock processes which our world does contain, such that θ_n is the time of state n, without specification however of the past-to-future direction of the series of states. The time interval to be associated with each state is arbitrary, and would in general depend on the cyclic clock processes that were being used and on the kind of physical measurements that were being made on each state. In some contexts a given state, of time θ_n, might be considered to endure for a year, while in other contexts a millionth of a second might be the suitable time interval for a state. We do exclude the possibility that the time of duration of a given state vanishes to a limit of zero, and we shall see in the discussion of the quantum theory in Chapter IX that there are good physical reasons for requiring a finite time interval for a state which is to be physically observed. It may seem to the reader that there is a difficulty arising from relativity theory in associating a unique time with each state; this matter will be discussed in Chapter VIII.

We can order our states in time, by writing the various θ's in an order, starting from the smallest and running to the largest, viz. θ_1, θ_2, . . . θ_n. The problem to be solved is, how do we determine the direction of time, from past to future, in this ordered series of states? The question may be more clearly realized by considering the "world" to be a limited domain, comprising a process of which motion pictures have been made. The states of the process are ordered in time by their positions on the picture film, but we may run the film in either direction and hence make two opposite interpretations of past-to-future succession in the process. For an organic process, as say the growth of a plant, it would be clear from our experience which are the earlier and later stages of the process, but we can think of a physical process, as say the rising and falling of a ball, for which the proper time direction of the process would not be so obvious.

In a series of states which are not all in the past or all in the future there are two groups of states, the past group and the future group, and also a single state, the present. The past-to-future direction is indicated by the present state, which does move in that direction, going continuously further away from a given past state, and continuously reaching future states which it then converts into past states. *In a given series of states,* in which each state differs from any other, the past group is the group which is always increasing in number, the future group is the group which is always decreasing in number. This statement of distinction is equivalent to saying that in a world which shows progressive change as ours does, the present state, as it becomes past, consistently adds itself as a new state to the group of past states.

The present state and the past states may be clearly distinguished from future states. The latter have not yet occurred, and hence cannot have been measured or observed with physical instruments as may be the present state or may have been the past states. Put in another way, in principle the state of any existing physical system may in some manner be recorded; hence the criterion for whether or not a state has occurred is whether or not there is an existent record of the state, or at least adequate reason for believing that there could be such a record. It is true that future states in a sense have existence: what they are to be may be predicted with varying degrees of accuracy. But future states of a system are not physically effective on a recording instrument which is set to respond to the states of a system contiguous with it, and this simple fact seems to be adequate in practice for distinguishing future states of a system from present and past states.

In order to find the actual sense of time direction in a set of past states that are ordered except for specification of time direction we appeal to the unique role of the state which is the "present state," the state which lies between future and past states. We do not need any unique present state; any one which is a present state for the investigator of the series of states will do. Suppose that θ_2 and θ_1 are two states of the series, indicated by their times, and we wish to know which of the two is the earlier state. We consider a present state, θ_p. If $|\theta_p-\theta_1|$ is greater than $|\theta_p-\theta_2|$, then θ_1 is earlier than θ_2, and the direction of time runs forward from θ_1 to θ_2. (The vertical lines indicate that the absolute value of the time interval, regardless of algebraic sign, must be taken.) Similarly, by reference to time interval with respect to a present state the earlier of any other two states among those which have occurred may be determined. Determination for any two will of course give the past-to-future time direction for the series.

Our criterion for determining the direction of time among a series of events or states is hardly an original one; in a word, we were merely saying that past to future direction of time is from things that happened long ago to things that happened not so long ago. Still, it is important to note the role of the present state, θ_p, in our determination of the time direction. Without a relation to θ_p we could not determine a time sense among a series of events; that is, we could not do so by the use of the cyclic processes (clocks) of our universe. In a series of recorded states, where θ_o is a state different from θ_p and θ_i any other state, the interval $|\theta_o-\theta_i|$ might refer to an interval that is "earlier" or "later" than θ_o. But, with θ_p, $|\theta_p-\theta_i|$ can only refer to an interval that precedes θ_p.

Our time concept, as built up in common experience, applies the earlier-later, or past-to-future, direction on events by virtue of the time interval of events from the present. Unless a set s of events or states are in a series of events that contains our own present state, we have no way of determining time direction among the states of s. All the events which are in our directed time series are, then, in our concept of time as derived from ordinary experience, part of a single stream or series of events. Events of a series which made no connection with the series of states that includes our present state—that is, which could not by use of our cycles be given θ numbers *in the series* including θ_p —would be "out of time" in the sense that we could not ascribe a past-to-future direction among them. This limitation on our assignment of time direction is, we will see, at least in part removed by the finding of relations between time direction and possible courses of physical process; specifically, by the relation between entropy and time-direction.

2. THE ENTROPY OF A SYSTEM.

One of our elementary natural experiences is that bodies which are hotter than their surroundings tend to become cooler, and similarly, cold bodies tend to take up heat from surrounding warmer bodies.

We learn that the rate at which heat differences tend to be equalized varies with different materials—the heat from a pot of boiling water diffuses out much more slowly through a wooden handle than a metal one—but in general heat always does move from a hot region to any cold region which is contiguous to it. Even a vacuum will not stop this movement of heat energy from hot to cold, although in a vacuum the transfer can occur only by radiation exchange, and not by direct molecular collisions as in matter of any kind.

The thermodynamic quantity, entropy, is defined with reference to the process of heat flow from one body to another. Rudolf Clausius introduced the essential elements of the entropy concept in a paper published in 1854, and in 1865 he proposed for the concept the term *entropy* from the Greek τροπη, *a turning in* or *transformation.* Later developments have shown entropy, in spite of its somewhat abstract quality as far as direct sense experience is concerned, to be one of the most useful conceptual inventions of science.

We define entropy primarily in terms of changes in the entropy of a body, with the use of certain stringent conditions of "reversibility" (to be discussed later) on the changes. If a body is at a uniform Kelvin (or absolute) temperature T, and if at that temperature the body gains a quantity of heat dQ, where dQ is a differential, so small that the temperature T is not changed by the quantity dQ, then we say the body has gained entropy dS, where $dS = dQ/T$. If the body has lost heat, $-dQ$, there has been entropy loss $dS = \frac{-dQ}{T}$. In order to determine the entropy change as a result of a total heat flow $Q = \int dQ$ we must integrate over the corresponding dS values. Thus, if "i" and "f" indicate initial and final states of the body, the entropy change in the body, ΔS, is $S_f - S_i$, and $\Delta S = S_f - S_i = \int_i^f \frac{dQ}{T}$.

In many processes T would be changing throughout the change from "i" to "f," and dQ and T must then be expressed in terms of a common variable which will be the variable of integration. There are, however, processes of heat flow and resulting change in which T is a constant. Thus, in the melting of a solid heat is supplied to bring about the change from "i" (solid) to "f" (liquid), with no consequent change in the temperature T of the material. The entropy change of the process is then simply $\Delta S = \frac{Q}{T}$, where Q is the total heat supplied to bring about melting and T is the temperature (the "melting point") at which melting occurs.

In order to define the entropy S of a body, rather than merely the change of entropy ΔS, we would have to begin the calculation of ΔS with an "i" in which the body possessed zero entropy. That is,

$S=\int_0^f \frac{dQ}{T}$, where "0" indicates the state of zero entropy. Customarily we assume that at zero temperature a simple chemical element has zero entropy content (the assumption is primarily just a convenient convention) and we may then assign an S value to a substance at temperature T if it has been possible to determine $\frac{dQ}{T}$ for the components of the substance over the path from zero temperature to the desired T. In practice physical chemists have been able to determine $\frac{dQ}{T}$ ratios very close to the absolute zero, and in tables of thermodynamic data one may find entropy values for many materials at stated temperatures.

3. ENTROPY AND NATURAL PROCESSES.

The observed transformations resulting from heat flow in nature occur in such a manner that the sum of the entropies of the bodies concerned increases as a result of the transformation. A simple heat flow situation illustrates the entropy increase. We think of two bodies, A and B, as being placed in contact. Initially the temperature of A, T_A, is greater than the temperature of B, T_B. The two bodies have no heat exchange, or other interaction, with any bodies other than themselves. Eventually A and B will come to a final state in which each body has the same temperature, but during the process of temperature equalization elements of heat, dQ, leave body A, of temperature T_A, and enter body B, at temperature T_B. Each heat loss dQ from A gives an entropy loss $dS=-\frac{dQ}{T_A}$ from A and an entropy gain $dS=\frac{dQ}{T_B}$ to B. Since $T_A>T_B$, the ratio $\frac{dQ}{T_A}$ is always greater than $\frac{dQ}{T_B}$ during the heat flow process, and there is a net gain of entropy by the two bodies as a result of the heat flow process which brings them to an equilibrium state in which they have equal temperatures.

No actual heat flow process in which there would be a decrease in the total entropy of the entities involved has ever been discovered or devised by physicists. Furthermore, by combining the first principle of heat flow (that heat by itself will flow only from hotter to colder bodies) with relations involving such properties of systems as pressure, volume, temperature, energy, and chemical composition, it has been possible to set up methods for calculation of entropy changes in a wide range of physical-chemical transformations, in principle, indeed, for any kind of natural transformation to which the variables of thermodynamics apply. These calculations have repeatedly been found to be valid, and physicists today have confidence in the generalization that all natural processes occur with either a net increase in entropy, or, just possibly, with no net entropy change; but never with a net entropy decrease.

The possibility of having transformations with no net entropy change is a theoretical one, and it arises with the concept of a "reversible" thermodynamic change. We return to the two bodies A and B, with a heat flow from A to B, to illustrate a reversible transformation with no net entropy change. Suppose that in some manner or other body A receives heat elements dQ, which flow from A to B. We require, however, that the temperatures T_A and T_B are virtually equal, so that heat dQ leaves A at temperature T_A and reaches B at the same temperature $T_B = T_A$. Then the entropy loss $dS = \frac{-dQ}{T_A}$ from A is equal and opposite to the entropy gain $\frac{dQ}{T_B}$ by B, and there is no net entropy change as regards the process of heat flow from A to B.

The definition of entropy change given on p. 21 required that in calculating entropy change between two states of a system we specify a hypothetical reversible change of state. We can make this requirement because entropy is a property of a given state, not of how the system actually reached that state. For a reversible process, any change of entropy in the system of interest is exactly compensated by an opposite entropy change in the surroundings; thus, in the reversible heat flow to A from B the entropy gain by A is compensated by entropy loss in B. Hence, for a reversible process the *net* entropy change in the universe is zero, although we find an entropy change for one component or system of the transformation process.

A heat situation between A and B such that heat elements dQ will flow from A to B, and yet with $T_A = T_B$, is frequently described as being one in which the heat flow takes an "infinitely" long time, for then at no time is there a measurable, finite temperature difference between A and B. A more meaningful characterization, in terms of what must occur physically, is that in the situation of heat flow without entropy increase there can be no disturbance from stable equilibrium states during the process of heat transfer. If at any time, A is hotter than B, there will be internal heat flow in A and B associated with the heat flow from A to B, since the process of transferring appreciable heat from A to B leads to internal temperature adjustments in A and B. But if the heat *flow* is occurring without entropy increase, there can be no temperature difference between A and B and hence also no differences internally in A and B. Hence, there is complete equilibrium as regards temperature at every stage of a heat flow process that is without accompanying entropy change. And, in general, transformation processes which do not involve an entropy increase may be characterized as processes which occur with no disturbance from equilibrium, or, in other words, which occur in a manner such that the system involved is at equilibrium at every state of the process.

A transformation which occurs with no net entropy change is referred to as "reversible" because it is one which may be conceived as being reversed in direction with no effects other than the reversal. There are no heat losses in friction or turbulence which involve departures from equilibrium and which cannot be reversed.

A reversible or no-net-entropy-change process is the most efficient kind of natural process, in the sense that there is less "waste," or redundant activity, than in entropy-increase processes. Any transformation which involves localized, irregular heating and heat flows, or frictional heat losses, or eddy current irregularities, or in general the kind of irregular processes that mean a disturbance from equilibrium states, is not a reversible transformation. Whether or not truly reversible thermodynamic processes occur in nature is a question open to speculation. Certainly some processes come close to the limit of being reversible processes, and it has been suggested that there are processes in living organisms, and also certain cosmic processes involving space expansion, that are reversible. There is today, however, no known process to which we can point as a clear illustration of reversible processes, and perhaps the weight of opinion among physicists is that the reversible thermodynamic process exists only as a never quite reached limit among natural processes.

Physicists have arrived, then, at the generalization that measurements of the entropy of the various components in any natural transformation will show that as the transformation proceeds the net entropy of the components will increase (or possibly, in certain limiting cases, remain the same). It is true that the entropy of every state is not necessarily determinable, for in general entropy is defined only for a state that is an equilibrium state; but the entropy of a known process can usually be considered in terms of an equivalent process which does proceed from an initial equilibrium state to a final equilibrium state, and this process has an entropy change value that is no greater than the true value for the process. For example, water flowing down a stream is not in an equilibrium state either at point *O* or at a point *P* downstream from *O*. But measurements would show that a mass of water, originally at rest in the stream at *O*, and allowed to flow to *P*, where it again was restrained and allowed to settle into an equilibrium state, would gain entropy in that flow process as a result of the frictional heat effects between the water and the confining stream bed. Likewise, it is certainly to be expected that water which is already flowing at *O*, and continues to flow at *P* will gain entropy as a result of the flow process from *O* to *P*. In general it has been found that analyses of natural processes into elementary processes for which entropy change estimates can be made give agreement with the generalization that natural processes occur with an associated entropy increase.

4. ENTROPY CHANGE AND TIME'S DIRECTION.

The problem of determining the past-to-future direction among a series of states of a system can be given a ready solution with the use of the entropy concept. *Successively later states of the system are states of the system with successively greater total entropy values.* The system must include all physically relevant components, for some one component of the system may very well lose entropy in its successive transformations, at the expense of greater entropy gains in other parts of the system with which the component is interacting.

A. S. Eddington[1] introduced the delightful notion of "Time's Arrow" with respect to entropy increase and time direction. The "arrow of time" moves forward in the direction of increasing entropy, and the direction of time is therefore simply along the course of natural transformations in the physical world. Because these transformations have the characteristic of so proceeding that entropy is always on the increase, time's direction is uniquely determined by reference to entropy change.

The association between increase of time and increase of entropy is not a primary definition of the forward direction of time. We have discussed earlier in this chapter how the past-to-future direction among a series of states may be determined in terms of time intervals from a present state; such a description of past-to-future relations seems to be not far from intuitive ideas of time's direction, ideas which are based on our experiences of a present state that itself successively becomes something different as it engenders newly formed past states. The notion of past-to-future direction rests primarily on such experience, or, more physically, on the changes that do occur in nature. In relating entropy and time we are making a genuinely synthetic statement: the direction of time, as observed in natural experiences generally, is such that later states of a complete system possess a larger value of the physical property, the entropy, than do earlier states. The entropy increase property is somthing that we learn about processes for which

[1]A. S. Eddington, *The Nature of the Physical World*, Macmillan, New York, 1929, Chap. IV.

we already well know the time relationships from time concepts that are firmly grounded in a wide realm of every-day experience.[2]

Once, however, the relationship has been established between the time direction of natural processes and entropy change we can, as with any natural law, use the relationship in situations in which one element of the relationship is unknown. By means of entropy considerations we can establish past-to-future time relations among a series of states of an isolated system which had no observed time relationship with our known sequence of time states. Thus, for instance, suppose an iron pillar, now no longer extant, to have been reported as once being of pure iron and at another time of rusted appearance. We know that iron and atmospheric oxygen form a state, with net increase in entropy, of iron rust, and we would not hesitate to place the iron bar in a rusty state as being of a later date than the bar in a pure state. We are constantly making temporal judgments on the basis of an associated physical process, and the generalization about entropy increase and time's moving arrow may be regarded as a statement about a property of processes, namely, entropy, and how that property changes as systems proceed from state to state.

It is noteworthy indeed that there is a property like entropy whose values may be so simply related to the forward direction of time. The entropy-time relationship has justifiably received wide attention. However, one important limitation on the relationship is that it does not tell us anything about the direction in time of individual atomic processes, for which entropy is not a relevant concept. Entropy is a thermodynamic variable, related by means of the equations of thermodynamics to other thermodynamic variables, as pressure, temperature, chemical composition, *et cetera.* All of these are "large-scale" variables, applicable only to systems containing many, many primary particles like electrons, protons, atoms, or molecules. Entropy considerations can tell us, then, about the time behavior of macroscopic systems, whose thermodynamic variables are descriptive of the average behavior of the entities that make up the system. But when we consider processes involving single particles, or only a few particles, we lose the guidance of the laws concerning entropy.

---

[2]We cannot agree with Reichenbach, *The Direction of Time,* California, 1956, pp. 125 to 135, who *defines* the direction of time in terms of entropy increase of the universe and its sub-systems. He thus explicitly concludes that time would be running backwards in a universe for which entropy were decreasing. We would urge, instead, that if the wider physical characteristics on which the time concept is based were present in such a universe, the expected entropy-time association would be that the past-to-future direction is associated with entropy decrease of a system. The view that entropy is the primary criterion of time's direction was effectively criticized by L. Susan Stebbing in her book, *Philosophy and the Physicists,* Methuen, London, 1937, chapter XI.

5. CYCLICAL PROCESSES: ENTROPY AND TIME.

The natural world of our direct experience is one of non-reversible processes and one for which entropy is therefore continually increasing. Nonetheless, there is an interesting theoretical point that emerges from a consideration of possible reversible processes and the role of cyclical phenomena in our concept of time.

Clausius' first generalizations about entropy change were made in connection with studies of an engine's going through a cyclical series of steps and converting heat into work, or vice versa. He concluded that at the end of a cycle, all steps of which had been carried out *reversibly,* the entropy of the engine and its surroundings would be just as at the beginning of the cycle, but if the engine cycle were carried out non-reversibly there would be a net increase in entropy at the end of each cycle. These conclusions may be generalized for any kind of defined cyclic process to which thermodynamic variables may be applied. As a very simple illustration we consider a pendulum bob, swinging in a vacuum. The motion of the pendulum would normally involve some friction at the point of suspension, and as a result of this friction heat production with concomitant entropy increase would occur in the surroundings (the suspension apparatus) during each cycle of the pendulum's motion. Hence, even though the "surroundings" might include a driving mechanism which maintained each cycle of the pendulum like its predecessors, the entropy of the system, pendulum plus surroundings, increases continuously during the pendulum motion. The direction of time could be determined by reference to the entropy of the system, a later pendulum swing being distinguished from an earlier one by the positive entropy production that had occurred in the complete system between the two swings.

In contrast, a hypothetical pendulum motion which proceeded reversibly, with elimination of all frictional heat losses, would involve no entropy increase in the pendulum or its surroundings. Hence, an attempt to determine in terms of entropy production in the pendulum system which of two pendulum swings was later in time would fail; with zero entropy production per cycle, there is no entropy criterion for the direction of time's arrow with respect to the pendulum swings.

The time relations of the frictionless pendulum swings could of course be readily determined by an observer, in terms of his own time sense. But it may be presumed that such an observer would himself be an entropy producing organism with his time sense capable of being correlated with the net entropy increase associated with his organism plus surroundings. Suppose, however, that the frictionless pendulum is part of a universe in which all processes are cyclical, with total absence of progressive, non-cyclical change in the universe.

There is then no process which is proceeding with net entropy increase and which provides an entropy measure for the past-to-future direction of time. The pendulum swings would thus be atemporal processes, and could not be put into any past-to-future time order.

We here find an interesting consonance between conclusions about time direction reached from general considerations of the time concept and from entropy considerations. We have seen in the previous chapter that if the natural world contained cyclical processes only we would not have gained our present time concept; instead, the world would be atemporal, since non-cyclical processes are necessary elements, in terms of which successive cyclic processes may be distinguished, in our concept of time. We get a similar result on entropy considerations; if the world consisted of purely cyclical processes, each cycle would have zero entropy change associated with it, since only with no net entropy change is a given cyclic process purely and completely cyclical, with the result that both the primary process *and its surroundings* return to an original state once in every cycle. But with zero entropy change associated with all natural processes, there would be no entropy change by means of which the direction of time could be fixed. The "philosophical" analysis in Chapter I—that is, an analysis of the elements of the time concept as we generally use and understand it—gave us the result that a world of cyclical processes only would not be a world containing a past-to-future time sense, and now our analysis of such a world of cyclical processes in terms of the relation between entropy and time gives us precisely the same result. The notion of a cyclical world's being an atemporal world therefore gains confirmation from the analysis based on the entropy concept. The agreement of results from the two kinds of analysis rests fundamentally on the fact that a world of cyclical processes each of zero entropy change is a world of no physical change, beyond that within each cycle, and likewise by direct definition the world of purely cyclical processes is a world of no progressive change.

6. TIME AND VARIOUS KINDS OF PROCESS.

We have found that the state of a total system which is undergoing transformations such that there is an associated entropy increase may be given a past-to-future time order on the basis of the entropy increase. The states of a total system which underwent cyclical processes only could not be put into a past-to-future time order; this conclusion follows either from the necessity of zero entropy change with purely cyclical processes or from our general conclusions about the elements required for a time order. A further hypothetical possibility would be a universe in which both cyclical and non-cyclical thermodynamic processes were occurring reversibly, and with no non-reversible processes, so that the total quantity of entropy remained unchanged. Such a universe would be without time direction on the basis of entropy-time associations, but presumably a time direction could be determined, since both elements that enter into the forming of our time concept, cyclical processes and progressive change, would be present.

Finally, there is the possibility of a universe all of whose processes are non-cyclical. Our general philosophical analysis has indicated that in such a universe there would not be a proper time concept, since there would probably not be regular processes in terms of which a uniform time interval could be measured. The processes of progressive change would however define a direction for time, so we might say that a "partial time" is defined by the non-cyclic processes, with regard to general experience with the processes. On the basis of entropy considerations, the universe of non-cyclic processes would give a measure of time's direction if the processes occurred non-reversibly, and would not give a measure if they were all reversible processes, occurring with no entropy change.

PROCESSES	Cyclical Only	Non-cyclical Only	Both Cyclical and Non-Cyclical
Reversible	Phil: *A* Entropy: *A*	Phil: partial *T* Entropy: *A*	Phil: *T* Entropy: *A*
Non-reversible	Non-existent	Phil: partial *T* Entropy: *T*	Phil: *T* Entropy: *T*

Kinds of Processes and Temporal Characteristics. "Phil" and "Entropy" indicate philosophical and entropy analysis respectively. *T*=temporal, *A*=atemporal.

These results for the temporal characteristics of various combinations of cyclical, non-cyclical, reversible, and non-reversible processes may be presented in the accompanying table. The processes are indicated as constituting a temporal or atemporal world, denoted by '*T*' or '*A*' respectively; assignment of *T* or *A* on the basis of our philosophical analysis of the time concept is indicated in the table by 'Phil,' while assignment on the basis of the entropy-time relation is indicated by 'Entropy.' It should be kept in mind that entropy considerations, as we have developed them, determine only the direction of time, so that 'Entropy: *T*' indicates only the determination of the directional element of time by entropy measurements. (It may be presumed, however, that cyclical processes, which served as units for time measurement in a world of non-cyclical and cyclical non-reversible processes, could be associated with entropy increases, each cycle of a regular process corresponding to a constant increase of entropy.) The temporal situation 'Phil: *T*,' which on the basis of the philosophical analysis is temporal, is one which possesses both of the elements required for our time concept. The situation which will be expected to yield a direction for time on the philosophical analysis, but no temporal measuring units, is indicated as 'Phil: partial *T*.'

The world all of whose processes are cyclical and non-reversible has been designated in the chart as "non-existent," for the following reason. A non-reversible cyclical process occurs with some net entropy change, that is, with some physical change in each cycle. Since the process itself is cyclical, the change must occur in the surroundings of the system that is undergoing the process. But if all the processes of the world were purely cyclical there would be no non-cyclical entropy change—or physical disarrangement—in the surroundings. Hence, there can be no non-reversible cyclical process in a world of cyclical processes only.

From the chart we see that the universe of reversible cyclical processes is at an extreme of atemporality; neither general experience nor entropy determinations would enable a time sense to be determined in this universe. For the other extreme we go to the lower right space of the chart; we see that we reach full temporality in a universe of both cyclical and non-cyclical non-reversible processes, with time units and direction determined on the basis of general experience and also time direction on entropy considerations. Our natural world appears clearly to be this kind of universe, although later discussions will give us grounds for at least entertaining the possibility of its being otherwise.

General References for Chapter II.

K. G. Denbigh, "Thermodynamics and the Subjective Sense of Time," *British Journ. Philos. of Science, 4,* 183 (1953).

A. R. Ubbelohde, *Time and Thermodynamics,* Oxford Univ. Press, 1947.

Adolf Grünbaum, "Time and Entropy," *American Scientist, 43,* 350 (1955).

Harold F. Blum, *Time's Arrow and Evolution,* Princeton Univ. Press, 1951.

R. Schlegel, "Time and Thermodynamics," in *Voices of Time,* ed. J. Frazer, Braziller, New York, 1966.

Chapter III

Reversible and Irreversible Processes

We have so far discussed the dependence of the time concept on reversible and irreversible processes, using the distinction that reversible processes are without any net entropy change in the universe whereas irreversible ones proceed with some uneven-ness or friction and hence do involve entropy increase. We have simply taken the two kinds of processes as given in nature, without further analysis of their composition. Now, however, we wish to take some account of the elementary processes which make up the observable processes of natural change. That is, we wish, as far as present day physical knowledge allows us, to go to the atomic level, and learn there what we can of time and the reversibility or irreversibility of natural process. In this and the following chapter we shall, then, be going from a level of complex thermodynamic processes, which involve countless millions of individual atoms or other entities, to a level at which individual atomic behavior is considered. The natural phenomena which we directly observe may be characterized as being "thermodynamic" in the sense that they involve "smoothed over" effects of many, many particles, or, equivalently, we may refer to these observed phenomena as being "large-scale" or "macroscopic." In contrast, consideration of individual atomic particles may be said to be of "elementary" or "microscopic" phenomena.

We shall see that the equations which govern atomic phenomena, on the microscopic level, are to a large extent equations which may be described as "time-reversible." Nonetheless, individual atomic processes do when combined together give large-scale effects which are irreversible. This fact constitutes one of the classic problems of physics, and in Chapter IV we shall discuss the solution that has been given for it. The existence of what appears to be widespread irreversibility in our world will then be treated in the light of what we know about reversibility in elementary and large-scale processes.

1. REVERSIBILITY PROPERTIES OF MECHANICAL SYSTEMS.

There are in physical theory numerous equations which have the properties that, a) they relate some physical quantity to a time variable, t; and, b) they have the same form whether t is present as a positive quantity, $+t$, or as a negative quantity, $-t$. These equations are equally valid, then, for time in the future as for time in the past, and we may refer to them or the processes which they describe as being "time-reversible."

A prime example of a time-reversible equation is afforded by Newton's second law of motion, which we write as

$$F_x = \frac{d}{dt}\left(m\frac{dx}{dt}\right), \tag{1}$$

where F_x is a force in the x-direction (which may be any chosen direction of space) acting on a mass m. The time differentiation is performed twice, and hence the differential elements dt which are involved in the differentiation may be either $+dt$ or $-dt$, without effect on the equation. This means that the equation equally well describes the effect of a force in producing accelerations of a mass m as positive time proceeds forward to the future or accelerations of the mass in time that is negative and is proceeding from the past.

We can perhaps better understand the time-reversible aspect of Newton's second law equation by considering a simple integrated form of it. Suppose that the force acting on the mass m is the force of

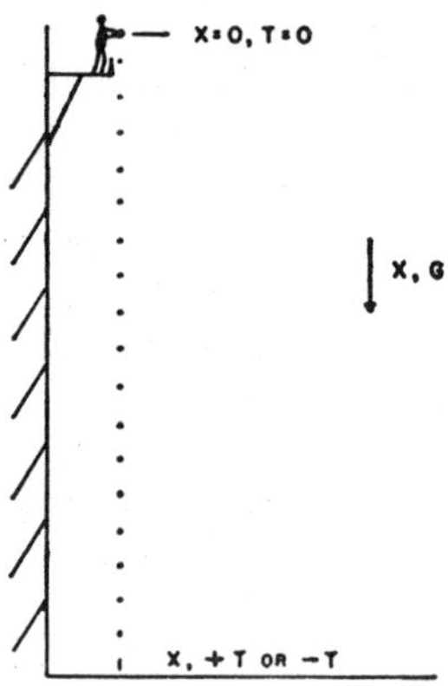

Fig. 1
The time-reversible equation describes the downward motion of the ball from $x=0$ to x in time from 0 to t, or the upward (reversed) motion starting at x at $-t$ and reaching $x=0$ at $t=0$.

gravity at the earth's surface, mg, where g is the measured acceleration of gravity. For simplicity we take the mass to be at rest, at the point $x=0$, at the time $t=0$, where now x is a vertical coordinate. Then, neglecting very high velocity relativistic effects which in any case do not explicitly involve t, the integrated equation gives us, for any mass whatever, $x=\frac{1}{2}gt^2$. This equation states a unique result for x, for either (+) or (−) t values. If, say, t=4, the equation states that $x=8g$ after 4 seconds; so if both g and x are taken as positive downwards, $8g$ will be the distance of fall at $t=4$. (See Fig. 1.) If $t=-4$, we also find $x=8g$. This means that if the mass is to be at $x=0$ at $t=0$, and all the while subjected to an acceleration g, it must be at $x=8g$ at $t=-4$. This second kind of motion is equally as "physical" as the simple falling of the mass from rest at $t=0$. Thus, if one wished to throw a ball upward, such that it would reach the height designated as $x=0$ at $t=0$, one would know that the ball must be at $x=8g$ at $t=-4$. A time differentiation of $x=\frac{1}{2}gt^2$ gives the velocity $\frac{dx}{dt}=gt$, and at $t=-4$ the the velocity must therefore be $-4g$; that is the ball must have upward (negative) velocity in contrast to the downward (positive) velocity of $4g$ which it would have at $t=+4$.

Or, we could solve the equation $x=\frac{1}{2}gt^2$ for t. We get two values, $t=+\sqrt{2x/g}$ and $t=-\sqrt{2x/g}$, and they give either future time, $+\sqrt{2x/g}$, at which time the ball will have fallen from $x=0$ to x, or the past time, $-\sqrt{2x/g}$, at which time it was at x on its rise to $x=0$. Solution for $\frac{dx}{dt}$ again indicates a positive velocity for the positive t value and a negative velocity for the negative value of t.

Another reversibility property of Newton's second law allows reversibility of a series of dynamical events such that with certain kinds of forces if a mass under the influence of the force moves from A to B, under the influence of the same force the mass may move from B to A. Both series of events, from A to B or B to A, are described by the same equation. Thus, for instance, we have seen that the motion of a ball subject to the earth's gravitational attraction may be either downward from $x=0$ to some other value of x or upward from that value of x to $x=0$. Which motion we have depends on the initial conditions: the falling in time t from 0 to x is reversed by giving the ball an upward velocity, $-\frac{dx}{dt}$, at x, instead of the falling downward velocity $+\frac{dx}{dt}$ which it would have after falling to x from 0. The upward moving ball has a final velocity, $\frac{dx}{dt}=0$, at $x=0$, and this zero velocity is the original velocity of the falling ball.

We may give the name "path reversibility" to the property of a time-reversible equation, whereby the setting of reversed final velocity as a new initial velocity gives rise to a retracing of the original motion. In order to gain further knowledge about path reversibility we shall now discuss it, on the basis of Eq. (1), using a more general mathematical treatment than in the preceding discussion.

Suppose that we have a position dependent force acting on the mass m. We write the force as $F(x)$, taking it, for simplicity, as a function of a single space variable, x. Eq. (1) then becomes

$F(x)=\frac{d}{dt}\left(m\frac{dx}{dt}\right)$, where again we take m to be a constant. Dividing through by m, and writing $\frac{F(x)}{m}$ as $a(x)$, we have

$$a(x)=\ddot{x}, \tag{2}$$

where the dots indicate time differentiation.

We integrate Eq. (2) between an initial point x_i and a final point $x=x_f$, and obtain

$$\dot{x}_f^2=2\int_{x_i}^{x_f} a(x)\,dx+\dot{x}_i^2. \tag{3}$$

Eq. (3), rewritten as

$$\dot{x}=\left[2\int_{x_i}^{x} a(x)\,dx+\dot{x}_i^2\right]^{\frac{1}{2}},$$

may in turn be written as if integrated between x_i and x_f. We obtain

$$\int_{x_i}^{x_f}\frac{dx}{\left[2\int_{x_i}^{x} a(x)\,dx+\dot{x}_i^2\right]^{\frac{1}{2}}}=t_f-t_i=t, \tag{4}$$

where $t_f=t$ and $t_i=0$. For the simple case of $a(x)=g$, a constant, we can see that Eq. (4) reduces to the equation, $x_f=\frac{1}{2}gt^2+\dot{x}_it+x_i$, which gives the position of a body that is freely-falling with constant acceleration g.

We apply Eq. (3) to a reversal of the motion from x_i to x_f; that is we reverse the velocity $\dot{x}_f$. We describe the reversed motion by primed symbols x_i', x_f', $\dot{x}_i'$, $\dot{x}_f'$, so that Eq. (3) becomes

$$\dot{x}'^2_f=2\int_{x'_i}^{x'_f} a(x)\,dx+\dot{x}'^2_i. \tag{3a}$$

By our condition of reversal, $x'_i=x_f$, $\dot{x}'^2_i=(-\dot{x}_f)^2$. These substitutions into Eq. (3a) give us

$\dot{x}'^2_f=2\int_{x_f}^{x'_f} a(x)\,dx+\dot{x}^2_f$, or using the value of $\dot{x}_f^2$ given in Eq. (3),

$$\dot{x}'^2_f=2\int_{x_f}^{x'_f} a(x)\,dx+2\int_{x_i}^{x_f} a(x)\,dx+\dot{x}_i^2. \tag{5}$$

We see that if $x'_f=x_i$, that is, if the reversed motion is allowed to carry the mass to the starting point of the original motion, Eq. (5) gives

$$\dot{x}'^2_f=\dot{x}^2_i, \tag{6}$$

which tells us that the reversed motion will have the same velocity magnitude as the original motion, since we may take x_i to be any point on the path of original motion. We can therefore say that $\dot{x}'^2=\dot{x}^2$, and also, that $\dot{x}'=-\dot{x}$. The latter equality follows because we also may reverse the velocity at any point; that is, any point on the original path may be taken as x_f. The path covered in time t in the original motion is $\int_0^t \dot{x}\,dt=x_f-x_i$. The path covered in the reversed motion, in an equal time interval t, is $\int_0^t \dot{x}'dt=x'_f-x'_i=\int_0^t -\dot{x}\,dt=-(x_f-x_i)$. Hence we have path reversibility: a particle with x_i, $\dot{x}_i$ at $t=0$ and x_f, $\dot{x}_f$ at t is given a reversed velocity $-\dot{x}_f$ at x_f, and will then return to x_i, $-\dot{x}_i$ after the same interval t.

Equation (4), which gives distance covered in time t, also displays path reversibility for explicit reversal of time. The equation may be rewritten:

$$\int_{x_f}^{x_i}\frac{dx}{\left[2\int_{x_i}^{x} a(x)\,dx+\dot{x}_i^2\right]^{\frac{1}{2}}}=-t. \tag{7}$$

Eq. (7) states that a distance from x_i to x_f, which is traversed in time t, will be traversed in reverse, from x_f to x_i as time goes from 0 to $-t$. In other words, Eq. (7) may be interpreted as stating that the particle, which moved from x_i to x_f in the time interval between $t=0$ and t, could by a "backwards" flow of time be brought from x_f to x_i in a time interval $-t$. Further, on differentiation of (7) with respect to t, using $d/dt=(dx/dt)d/dx$ and taking the limits to be variables, for the integral, we find:

$$\frac{dx}{dt} = -\left[2\int_{x_i}^{x} a(x)\,dx + \dot{x}_i^2\right]^{\frac{1}{2}}. \tag{7a}$$

The velocity dx/dt at any point in the backwards motion is opposite, then, to the velocity

$$\dot{x} = \left[2\int_{x_i}^{x} a(x)\,dx + \dot{x}_i^2\right]^{\frac{1}{2}}$$

that we found for the forward motion. (The velocity could of course originally be given a positive or negative value, depending on whether a (+) or a (−) sign is assigned to the square root in (4) and (7); but the velocity sign obtained from Eq. (7a) would in any case be the negative of that assigned for the original motion.) The reversal of path, as a result of velocity *and* time reversal, is shown by Eq. (7) to be mathematically possible, and there perhaps are physical situations in which it is proper to speak of a reversal of time direction, as will be discussed in section 3 of Chapter VI. One might also consider a path reversibility in which there is no velocity reversal but only a change in direction of time; the difficulties in connection with this kind of reversibility will be discussed with Eqs. (15)–(18) of this section.

Many of the forces which operate in nature are of the form $F(x)$; Newtonian gravitational forces, which vary with distance from a central body, are an example. With such forces we have, then, complete path reversibility, as displayed by Eqs. (6) and (7): reversal of velocity gives reversal of path, and also reversal of time gives reversal of velocity and of path. It must not be thought, however, that an equation like (1), which as a differential equation is time-reversible, will always give path reversible equations of motion on integration. The integration of (1) with a time dependent force, $F(t)$, gives an instance of an equation of motion gained from (1) which does not possess the property of path reversibility, as we shall now show.

We again take m to be constant, and write $F(t)/m = a(t)$. Then we may write from (1),

$$\ddot{x} = a(t). \tag{8}$$

We assume $a(t)$ to be an "even" function of t, such that $a(t) = a(-t)$. The first integration gives:

$$\int_{\dot{x}_i}^{\dot{x}_f} d\dot{x} = \int_0^{t_f} a(t)\,dt, \text{ or } \dot{x}_f = \int_0^{t_f} a(t)\,dt + \dot{x}_i, \tag{9}$$

where we take $\dot{x} = \dot{x}_i$ at $t = 0$ and $\dot{x} = \dot{x}_f$ at $t = t_f$. For t_f taken to be a variable t, Eq. (9) is

$$\dot{x}=\int_0^t a(t)\,dt+\dot{x}_i. \tag{10}$$

Integration of (10) gives

$$x_f=\int_0^{t_f}\left[\int_0^t a(t)\,dt+\dot{x}_i\right]dt+x_i, \tag{11}$$

where $x=x_i$ at $t=0$ and $x=x_f$ *at* $t=t_f$.

We consider a reverse path, with the initial and final points and velocities again symbolized as x_i', x_f', $\dot{x}_i'$, $\dot{x}_f'$. Eq. (10) for the reversed motion and for the same time interval as before is:

$$\dot{x}'=\int_0^t a(t)\,dt+\dot{x}_i'. \tag{10a}$$

Since $\dot{x}_i'=-\dot{x}_f$, we have, using (9),

$$\dot{x}'=\int_0^t a(t)\,dt-\int_0^{t_f} a(t)\,dt-\dot{x}_i. \tag{12}$$

Eq. (12) tells us that $\dot{x}_f'$, which is $\dot{x}'$ after a time t_f, is simply $-\dot{x}_i$. Hence, in time t_f, starting with the reverse of the first final velocity, we will come to a reverse of the first initial velocity. This seems to be a satisfactory reversibility situation. However, we do not find path reversibility from Eq. (12). Integration of (12) between the limits x_i' and x_f' gives us:

$$\int_{x'_i}^{x'_f} dx=\int_0^{t_f}\left[\int_0^t a(t)\,dt-\int_0^{t_f} a(t)\,dt-\dot{x}_i\right]dt, \text{ or,}$$

$$x_f'=\int_0^{t_f}\left[\int_0^t a(t)\,dt-\int_0^{t_f} a(t)\,dt-\dot{x}_i\right]dt+x_i'. \tag{13}$$

We rewrite (13), using $x_i'=x_f$, and Eq. (11) for x_f. We then have:

$$x_f'=\int_0^{t_f}\left[\int_0^t a(t)\,dt-\int_0^{t_f} a(t)\,dt-\dot{x}_i\right]dt$$

$$+\int_0^{t_f}\left[\int_0^t a(t)\,dt+\dot{x}_i\right]dt+x_i, \text{ or,}$$

$$x_f'=\int_0^{t_f}\left[2\int_0^t a(t)\,dt-\int_0^{t_f} a(t)\,dt\right]dt+x_i. \tag{14}$$

We see from Eq. (14) that in general the particle, after a velocity reversal at x_f, will not return to x_i in the time t_f in which it moved from x_i *to* x_f. Instead, after a time t_f the particle will be at x_f' as given by (14). Hence, we do not have path reversibility. In the special situation in which $a(t)$ degenerates into a constant acceleration, g, we see, from Eq. (14), that

$x_f' = \int_0^{t_f} [2gt - gt_f]dt + x_i = x_i$, and we do have a return to the initial point in time t_f.

An $F(t)$ which varies sinusoidally, so that $a = A_x \cdot \sin \omega t$, where ω is an angular frequency, and A_x an amplitude factor, provides an interesting example of non-reversibility. Such a force is applied for the same length of time, and with equal magnitudes, in both the $+x$ and $-x$ directions; yet, integration of Eq. (1), with the acceleration as proposed, shows that a particle originally at rest is given an increasingly larger displacement as t increases from $t=0$.[1] That is, there is no path reversibility, although the sign of the applied force changes regularly as t increases.

Even though we do have path-reversibility for the basic equations of Newtonian mechanics, David Finkelstein[2] has shown that the gravitational field of a spherical particle, as described by the results of general relativity theory, is not invariant under a time-reversal. Also, an acceleration which may well be a function of time and which appears to be important in the physical world is the mutual expansion acceleration of the galaxies (to be discussed in Chapter V). There may be other such accelerations, but we again note that most physical forces with which we have experience are "field" forces that are expressed as $F(x)$.

It is instructive to see that we do get path reversibility with an $F(t)$ —in contrast to the result of Eq. (14)—if we mathematically move "backwards" through time. Instead of carrying out reverse motion by a reversal of velocity, we integrate in time through negative time differentials, $-dt$. We start the motion at time t_f, with the moving mass m at x_f.

Equation (1), written with negative increments $-dt$, for $\frac{F(t)}{m} = a(t)$, is,

$$\frac{-d}{dt}\left(\frac{-dx'}{dt}\right) = a(t). \tag{15}$$

A first integration of (15) gives

$$\frac{-dx'}{dt} = -\int_{t_f}^{t} a(t)\,dt + \text{const.} \tag{16}$$

---

[1]G. Joos, *Theoretical Physics*, trans. by Ira M. Freeman, Second Edition, Hafner, New York, prob. 31, p. 83.

[2]"Past-Future Asymmetry of the Gravitational Field of a Point Particle," *Physical Review, 110*, 965 (1958).

In (16), t_f is the final time, at which the motion described by Eqs. (9)–(11) has come to an end; t is some earlier time. At $t=t_f$, $\dot{x}'=\dot{x}_i'=\dot{x}_f$, so the const. of (16) is $-\dot{x}_f$. Hence we write (16) as:

$$\dot{x}'=\int_{t_f}^{t} a(t)\,dt+\dot{x}_f.$$

By substitution of the value of $\dot{x}_f$ as given by (9),

$$\dot{x}'=\int_{t_f}^{t} a(t)\,dt+\int_{o}^{t_f} a(t)\,dt+\dot{x}_i=\int_{o}^{t} a(t)\,dt+\dot{x}_i. \tag{17}$$

Integration of (17), between the limits x_i' at $t=t_f$ and x_f' at $t=0$ gives:

$$x_f'=\int_{t_f}^{o}\left[\int_{o}^{t} a(t)\,dt+\dot{x}_i\right]dt+x_i'.$$

Since $x_i'=x_f$, we find, using (11):

$$x_f'=\int_{t_f}^{o}\left[\int_{o}^{t} a(t)\,dt+\dot{x}_i\right]dt+\int_{o}^{t_f}\left[\int_{o}^{t} a(t)\,dt+\dot{x}_i\right]dt+x_i,$$

or,

$$x_f'=x_i. \tag{18}$$

Equation (18) assures us that a hypothetical complete reversal of physical motion by means of a "going backwards" of time is a mathematical possibility.[3] Such a reversal, however, has little if any physical plausibility. We see from Eq. (17) that the particle, which begins its return motion at t_f, would at t_f have a velocity $\dot{x}_i'$ that is equal to the velocity $\dot{x}_f$ (Eq. (9)) with which the particle ended its original motion. Likewise, at other comparable times on its return the particle would have the same velocity, in magnitude *and* sign, as it had for the original motion. Finally, at $t=0$, we find $\dot{x}'=\dot{x}_i$, which is the original initial velocity specified by Eq. (9). We have, then, the remarkable physical situation that in its motion with "reversed time" but no velocity reversal the particle at all points has a velocity that is carrying it in a direction from x_i to x_f, and yet successive positions of the particle, as it goes through $-dt$ increments, are successively closer to x_i and farther from x_f. Such a motion is physically contradictory unless one invoke

[3]Another possibility is to carry out a reverse motion by reversing both velocity and time. We have seen in Eq. (7) that with an $F(x)$ such a reversed motion gives path reversibility. In the case of an $F(t)$ we do not however get path reversibility with such reversed motion. The interested reader can show that if integration is carried through, "backwards" in time but with $\dot{x}_i'=-\dot{x}_f$, the result corresponding to Eq. (18) is

$$x_f'=\int_{o}^{t_f}\left[2\dot{x}_i+\int_{o}^{t_f} a(t)\,dt\right]dt+x_i.$$

some kind of highly contrived sub-stratum temporal flow which carries physical events with it. Thus, analogically we might think of time's flow as a great conveyor belt, on the surface of which physical events occur. A particle moving in one direction on the belt might be carried in an opposite direction by a reversal of motion—that is, of time's direction—of the entire belt. We reject, however, any such sub-stratum concept of time, on the grounds that it goes beyond physical theory and observation. We conclude, therefore, that the motion described by a mathematical reversal or retracing of time only is physically contradictory.

We saw in Eq. (7), when working with an $F(x)$, that literal time reversal brought path-reversibility, but with an associated velocity reversal. In that equation we have an explicit relation between x and t, and the physically expected result of velocity reversal with time reversal is required by the equation.

In general it is not possible to retrace motion *in every respect* by going "backwards in time" because the nature of a reverse process prevents complete reversal of a process. A simple time-reversible system, let us say, starts at an initial configuration, I, and goes through various motions, eventually reaching a final configuration F. (The system might be as simple as the movement of a mass from an initial point x_i to a final point x_f.) By reversing the final velocities we send the system through the reverse path, from F to I, exactly retracing the path that led from I to F. At any point between F and I the system will be in a configuration identical to the corresponding configuration as it moved from I to F, except that in the reverse motion all velocities will be the opposite of what they were in the original motion. It is then impossible to have a reversal in which there is complete identity, even to equal velocity directions, between the original and reverse paths, for by definition the reverse motion requires a different direction of motion than the original motion. Indeed, should a system go from configuration I to F, and then go through the same series of configurations, even with identical velocity directions, the system then has not reversed but has *repeated* itself. It is of course a trivial property of any physical equation that it permits repetition of a series of events, from the same initial conditions.

In summary, we have seen in this section that a time-reversible equation like (1) is invariant for the transformation from (+) to (−) values of the time variable. We considered the reversibility properties of solutions of such an equation under three different conditions: i) reversal of velocities but usual "forward" flow of time; ii) reversal of velocities and also reversal of time; iii) reversal of time only. We saw, Eqs. (2)–(6), that conditions (i) give us path reversibility for forces that are functions of position, $F(x)$; but (Eqs. (8)–(14)) we do not find such reversibility for a force that is an arbitrary function of the time. With conditions (ii), we have path reversibility for $F(x)$, as seen in the discussion of Eq. (7), but (Footnote 3) again not for $F(t)$. Conditions (iii) will give a formal retracing of the original path with an $F(t)$: see Eqs. (15)–(18). The implied motion, however, with a "backwards" flow of time but no velocity change, is physically unreasonable.

2. REVERSIBILITY AND MAXWELL'S EQUATIONS.

Maxwell's four equations of electromagnetism may be regarded as stating the fundamental properties of electrical and magnetic fields, just as Newton's second law states the basic relation between force and motion. These equations likewise are time-reversible, although in a somewhat different manner than are the basic mechanical equations.[4]

The first two of Maxwell's equations are for a "steady state," in which no change is occurring, and include no reference to time. Thus, the equation $\nabla \cdot \mathbf{E}=4\pi\rho$, where **E** is electric field intensity in a vacuum and ρ is charge per unit volume, may be called a time-less equation; it tells us what **E** will be, in the space where we have a given charge distribution, but it says nothing about how **E** will change as ρ is changing. Similarly, the second equation, $\nabla \cdot \mathbf{H}=0$, where **H** is magnetic field strength in a vacuum, is without time implication.

The remaining two equations do involve reference to state of change, and, specifically, they both explicitly refer to time rate of change. We consider the simpler of the two equations, that is,

$$\nabla \times \mathbf{E} = -\frac{1}{c}\frac{\partial \mathbf{H}}{\partial t}, \tag{1}$$

where t is the time variable and c is the velocity of light. It has an appearance of being a time-irreversible equation, in that a substitution of $-t$ for t in the equation would give a change of sign of $\nabla \times \mathbf{E}$. This change of sign does in fact guarantee, however, that a process determined by an **E** field set up in accordance with Eq. (1) will retrace its path if t is reversed, much as the change of sign of velocity in reversible mechanical equations gives path reversibility. Thus, let us consider an integral equation which is equivalent to Eq. (1):

$$\oint_C \mathbf{E} \cdot ds = -\int_S \frac{\partial \mathbf{H}}{\partial t} \cdot dS. \tag{2}$$

In Eq. (2) ds is an element of curve C bounding the surface S. The

[4]A discussion of reversibility of physical laws, much more general and advanced than the one presented in this chapter, is given by Satosi Watanabe, *Rev. Mod. Physics, 27,* 26 (1955).

equation tells us that the normal component of the time rate of change of **H**, taken over the surface *S*, gives rise to an electric potential difference $\mathbf{E} \cdot ds$ in the bounding curve of the surface. This potential difference—or, the field **E** associated with it—would, for instance, give rise to motions of electrons or ions in a conductor that formed the boundary curve *C*. If the motions resulting from the applied electrical force were themselves time-reversible, that is, as described by a time-reversible mechanical equation, then a reversal of the sequence of electromagnetic events would result in an exact reversal of the charged particle velocities. For, if originally **H** increased from $\mathbf{H}_i$ to $\mathbf{H}_f$, a reversal from $\mathbf{H}_f$ to $\mathbf{H}_i$, giving the reverse succession of $\frac{\partial \mathbf{H}}{\partial t}$ magnitudes, would give a reverse succession of electrical forces; but now each $\frac{\partial \mathbf{H}}{\partial t}$ would have a negative sign, because **H** would be decreasing, and the sign of **E** throughout would change. Hence, the electric forces would be reversed in sign, and there would be a reversal of the original velocities. Or, alternatively, we could think of the original increase of **H** from $\mathbf{H}_i$ to $\mathbf{H}_f$, in a time from 0 to *t*, as being reversed by a hypothetical backwards motion of time, from *t* to 0. Again, the sign of each $\frac{\partial \mathbf{H}}{\partial t}$ magnitude would change, because of the negative *t* increments, and hence the sign of **E** would change, thereby giving the reversal of particle velocities.

An equation like (2) will lead, then, to velocity reversibility in the manner that has been specified. Eq. (2) has this property because a change of direction of the induced electrical field is concomitant with a reversal of the time succession of magnetic states. We do not get path reversibility with simple velocity reversal of a charged particle, if the particle is moving in a magnetic field and interacting with the field as a result of its motion. Suppose the particle is moving with a velocity *v* in a plane normal to a uniform magnetic field **H**, so that the resulting force *F* causes the particle to describe an arc of a circle of radius *R*, in a counter-clockwise sense, about a center *O*. If the velocity of the particle is reversed, the direction of the force on it will also be reversed in accordance with the well-known rules of charge-field interaction: the particle, instead of retracing the arc path on which it has moved, will move in the same sense about a new center removed from *O* by a distance 2*R*. (See Fig. 2.) On the other hand, if the velocity reversal of the particle is part of a velocity reversal of all particles (as would be expected with a time reversal), we do gain path reversibility. The motions of the charged particles which are producing the magnetic field would then be reversed, and there would be a resulting reversal of the sense of the magnetic field. With a reversal in the directions of both the

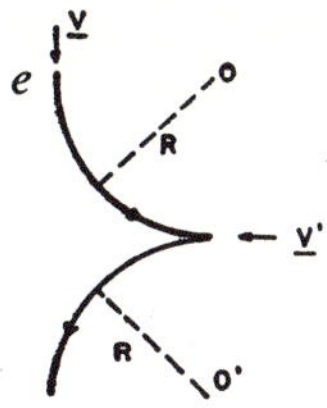

Fig. 2
H is uniform and perpendicular to the plane of the sketch. The charged particle *e*, with an original linear velocity v, moves in the circle of radius *R* about *O*. When the velocity of *e* is reversed, to v′, the particle describes a new arc, along a circle of radius *R* about *O′*.

magnetic field and the velocity of the moving test particle, the direction of the force on that particle would not change. This instance of motion with electromagnetic forces, and the one discussed in connection with Eq. (2), suffice to show that within electromagnetic situations one cannot make simple generalizations about reversibility, but must consider various types of different situations.

3. IRREVERSIBILITY OF MANY-PARTICLE PROCESSES.

The equations which we have been discussing are very simple ones, in that they relate forces and motions for a mass taken as a unit, or electrical or magnetic fields all parts of which are described by a single field vector. The equations do not, for example, describe phenomena which involve an average effect of many kinds of motion. There are phenomena in nature which are described to a high degree of accuracy by an equation for a simple process; the motion of a planet around the sun, for example. Most of the natural processes which we observe are, however, average effects of many individual processes. Further, these natural processes proceed with entropy increase and therefore are irreversible: the processes cannot by a reversal of conditions be made exactly to retrace themselves. Thus, a gas which expands from a small container into a large one shows effects of the motions of the many individual molecules which constitute the gas, and one cannot reflect the molecules, so that they reverse their velocities and return to the smaller container.

An example of equations for irreversible processes, as contrasted with the equations that permit path-reversible solutions, is provided by the diffusion equation,

$$\frac{\partial C}{\partial t} = D\frac{\partial^2 C}{\partial x^2}. \qquad (1)$$

This equation governs the variation in concentration C in time t, when two media of different C values are brought into contact in a plane normal to the x-axis. The C may represent solute in a solvent, momentum of particles, or a temperature value. D is an appropriate "diffusion constant."

The general solution of (1) is of the form:

$$C(x,t) = e^{-k^2Dt}(ae^{ikx} + be^{-ikx}), \qquad (2)$$

where k, a, and b are constants which are determined by the boundary conditions of the specific physical situation. Suppose now that we consider the change in C at a fixed x value, x_0, assuming that $C = C_o$ at $t=0$. Then, we find from (2) that C_o at x_o is equal to $ae^{ikx_0} - be^{-ikx_0}$. If k is a real number, C will decrease exponentially as time goes on, since now $C = C_o e^{-k^2Dt}$. For example, at a t such that $-k^2Dt = -5$, we find $C = 0.0067C_o$. Suppose, in contrast, that we go from $t=0$ to a negative t value. We then have an increasing exponential behavior of C, which is quite different from the decreasing exponential behavior: with a change in sign of t, so that $-k^2Dt = +5$, we find $C = 148.4C_o$. Clearly, if we first change from C_o to C_1 in a time interval of, say, one second, our equation would not predict a value of C_o if we then went through a time interval of a negative second, starting with C_1 at $t=0$. Diffusion processes are irreversible physically, in that they cannot be reversed in physical fact, and the equations describing diffusion are time irreversible because the curve for C, with increasing t, is different from the curve with decreasing t. The difference between the exponential curves for (+) and (−) values is illustrated in Fig. 3. We see

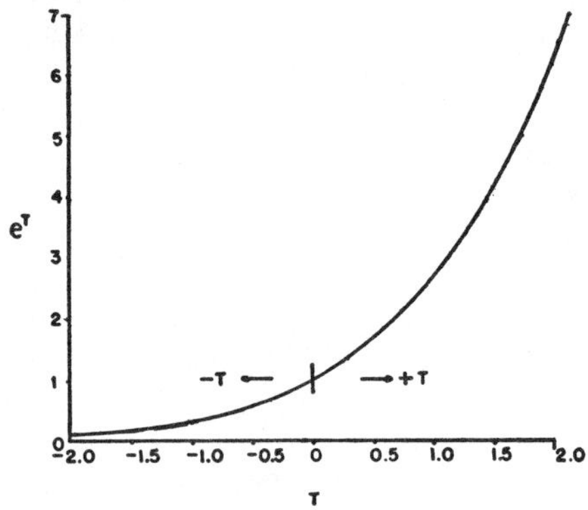

Fig. 3
The term e^t is plotted for (+) and (−) values of t.

that the curve for negative t in no way reproduces the curve for positive t values.

The equation for a wave amplitude, u, varying in time t along a coordinate x, has the form,

$$\frac{\partial^2 u}{\partial t^2} = v^2 \frac{\partial^2 u}{\partial x^2}, \quad (3)$$

where v is the wave velocity, here taken to be a constant. The second time-derivative in (3), as compared with the first derivative in (1), converts (3) into a time-reversible equation. It is clear that the wave equation is unchanged by the substitution of $-t$ for t. Further, the solution of (3) may be given in the form,

$$u = a \cos (x \pm vt) + b \sin (x \pm vt). \quad (4)$$

In (4), a and b are arbitrary amplitude factors, and the sign between 'x' and 'vt' is $(+)$ for a wave that is travelling in the negative x direction, with speed v, and $(-)$ for one that is moving with that speed in the positive x direction. We see then, that the waves which are described by solutions of (3) will in general have properties of path-reversibility, such that $+t$ intervals describe the wave moving in one direction and $-t$ intervals describe the wave moving, otherwise identically, in the opposite direction.

The entropy increase which is associated with many-particle processes requires that these be irreversible processes. The term "dissipative processes" is often used for irreversible processes; they occur with energy dissipation—that is, with frictional or other loss of energy. The processes cannot be reversed because the random motions will not be reversed in the reversal of the major trend of the process. Thus, when a wave is set up in a string, in accordance with Eq. (3), the wave motion is not exactly as described by (3), and is not strictly a reversible process. There are frictional energy losses, both at the string mounting and within the material of the string, which cause successive waves at a given point to decrease in amplitude (assuming that there is no energy source which is maintaining the wave amplitude). Thus the vibrations of a plucked violin string do eventually die away.

The oscillations of a light wave moving through a vacuum are also described by Eq. (3), and in this case there is no dissipation—at least, we have no reason to believe that a photon of light, even in coming from the far galaxies, suffers any loss of energy as a result of its travelling through space. The light wave in a vacuum is, then, a reversible process, and by mirror-reflection light may be made to retrace a given path. The passage of a photon through space is not however a process to which we can assign an entropy, for it is not a many-particle process which involves average effects of many entities; rather, it is a single entity process, accurately described by a time-reversible equation.

General References for Chapter III.

R. C. Tolman, *Principles of Statistical Mechanics,* Oxford Univ. Press, 1938, secs. 37 and 95.

H. Reichenbach, *The Direction of Time,* U. of California Press, Berkeley, 1956, secs. 4, 7, 24.

P. W. Bridgman, *The Nature of Thermodynamics,* Harvard Univ. Press, 1943, p. 125 *et. seq.*

G. Morpurgo, B. F. Touschek, and L. A. Radicati, "On Time Reversal," *Il Nuovo Cimento, 12,* 677 (1954).

Discussion of Irreversibility and Mechanics, K. R. Popper and R. Schlegel, *Nature, 178,* 381 (1956); K. R. Popper, E. L. Hill, and A. Grünbaum, *idem., 179,* 1296 (1957); R. C. L. Bosworth and K. R. Popper, *idem., 181,* 402 (1958).

Chapter IV
H-Theorem and Entropy

1. BOLTZMANN'S H-THEOREM.

We have seen that there are differential equations which are irreversible in the time variable, and also there are those which are reversible. The latter will in general admit both kinds of solutions: those which display path reversibility and those which do not. There is a very large and important class of path-reversible solutions of the fundamental equations of mechanics; indeed, in general all solutions of Newton's second law in which the force is not a function of time have the reversibility property. In most uses of Newton's laws, force is a function of position; the non-reversible solutions with $F(t)$ which we have discussed in the previous chapter have not been of great importance in physics. There are, then, many single-element processes of nature which are described by equations with the property of complete path reversibility.

A fundamental problem of physics arises when we consider the single-entity, reversible processes as elements in many-entity processes. Observation has always confirmed that these latter processes occur in nature with entropy increase; that is, they are irreversible processes. *How can completely reversible processes combine to form a process that is irreversible?* If the individual constituent processes should each be reversed, would not the larger process of necessity be likewise reversed? This difficult problem is answered by the H-theorem of Ludwig Boltzmann.

The expansion of an ideal gas provides a simple illustration of the problem which is solved by the H-theorem. The expansion, say from a container A into a larger container B, originally empty, is clearly an irreversible process. The molecules of the gas, however, undergo elastic collisions with each other, and there is no energy loss in these collisions. The equations of motion of the molecules, in such a case where total energy is conserved, must have the property of path reversibility. Indeed, we can neglect any consideration of what occurs during collisions, and only consider each molecule as having some constant velocity at any one instant; that is, we consider the molecules as they are moving between collisions, with the proviso that the changes of velocity suffered at collisions are such as to maintain constant total energy of the gas (in accordance with observation). For a molecule at constant velocity, Eqs. (12) and (14) of section 1 of chapter III may be applied, with $a(t)=0$; the molecules obviously then have the property of path reversibility. We ask, since it is a matter of indifference whether a molecule have a given velocity or its reverse, why should not a process like gas expansion equally well occur in reverse, that is, why should not gas contraction be as physically likely as gas expansion?

In considering the gas expansion we must think in terms of the physical laws governing the individual molecules, if we are to appreciate the problem involved. We well know that a gas does expand into a region of lower pressure, and physical law describing the behavior of gases reflects this knowledge. But if we forget, as it were, that we know anything about how gases behave in the large, and if we consider only the individual molecules of a gas, moving in accordance with the (reversible) laws of motion, we see no reason why a gas contraction should not be as physically plausible as a gas expansion. What we must discern is, what is the factor on the level of molecular behavior, additional to the laws of motion, which determines that the gas expands rather than contracts when placed in contact with a region of pressure that is lower than the original gas pressure. An additional "irreversibility factor" must enter into the situation, since the laws of motion of the individual particles are completely reversible.

Boltzmann's work shows in effect that the "irreversibility factor" is a statistical consideration which becomes important when large numbers of entities are involved in a process. He considers a gas whose molecules are described by reversible dynamical equations. Further, there exists for these molecules a velocity distribution function F which is a function of the position coordinates, x, y, z, and of the momentum coordinates p_x, p_y, p_z of the molecules, such that $dN = F\,dx\,dy\,dz\,dp_x dp_y dp_z$ gives the number of molecules to be found in the "volume" $d\omega = dx\,dy\,dz\,dp_x dp_y dp_z$. A quantity H is defined:

$$H = \int F \log F d\omega,$$

where integration is carried over the entire range of the x,y,z,p_x,p_y,p_z coordinates.

Boltzmann's H-theorem states that dH/dt, the time derivative of the quantity H, will never take a positive value. That is, $dH/dt \leqslant 0$. The quantity H, therefore, will never increase, but as time goes on will decrease or remain constant. Further, the equations that lead to the H-theorem require that at an equilibrium state of the gas $dH/dt=0$. The theorem tells us, then, that the gaseous system approaches equilibrium as time goes on, and that once it reaches equilibrium it remains in that state. Also, at equilibrium, it can be shown that the distribution function *F* is such that the gas takes on a uniform density in any container. Hence, the *H*-theorem requires that a gas will expand so as uniformly to fill an increased volume made available to it.

We shall not here present a derivation of the *H*-theorem; the interested reader can find such in a treatise on statistical mechanics. But we point out that the *H*-theorem rests on assumptions of conservation of energy and of conservation of momentum in the molecular collisions in the gas, and also on an assumption of complete randomness or *molecular chaos,* as it is sometimes called, in the intermolecular relations. This last assumption specifies that there will be no correlation between the velocity of a molecule and the kind of molecules that are around it; loosely speaking, we might for example say that the assumption forbids "bunching" such that the fast-moving or slow-moving molecules are grouped together. Another way of stating the assumption is to say that it requires, on the average, that each molecule regardless of its velocity meet the same distribution of molecular velocities in its collisions as does any other molecule. The *molecular chaos* assumption brings a basic statistical element into the *H*-theorem, since the assumption is meaningful only in terms of the existence of a large number of molecules, and is not one for which there is an expectation of exact and rigorous validity, as with the assumptions of energy and momentum conservation. This statistical element is of importance in considering objections to the *H*-theorem, as we shall see in the next section.

The *H*-theorem is a keystone in our understanding of natural process. The irreversibility of those natural processes for which the constituent processes are reversible would be mysterious indeed without the theorem. Having the theorem, we can see at least for a few representative processes—subject to the assumptions that we have noted—that collisions among a large random group of molecular elements will give the group that behavior which is observed in nature. For, granted its assumptions, the *H*-theorem shows with mathematical rigor that the inter-molecular collisions of a gas can lead only to the states of the gas, of decreased H, that are reached in observed changes. The "irreversibility factor" that prevents large-scale reversible changes arises from the consideration of the *many* interactions of the molecules in the gas. The molecular collisions which send molecules from one momentum state to another—and hence bring changes in the *F* function which is involved in *H*—can act only to decrease *H* or keep it constant, and so, with consideration of many interactions, individual reversibility is converted into group irreversibilty.

Why this conversion occurs is not intuitively obvious, even with study of the details of the H-theorem derivations. A discussion of two well-known objections to the H-theorem, to be presented in the next section, will however give us some insight into the nature of the irreversibility that the H-theorem expresses.

2. ARGUMENTS AGAINST THE H-THEOREM.

There are two noteworthy arguments against the H-theorem. One, first given by J. Loschmidt, rests on a hypothetical reversal of molecular velocities in a gas. The second, given by E. Zermelo, involves the presumed repetition, in a sufficient length of time, of any given molecular configuration. The two objections are, in the tongue of their authors, the *Umkehreinwand* and the *Wiederkehreinwand.*

Loschmidt's objection has been tacitly discussed in the previous section, in the mention of the possibility of reversing all the velocities of the elements of a system. Because of the path reversibility of the equations governing the molecules of a gas, such a reversal is physically possible, and the large-scale behavior of the gas would be turned about as a result of such a velocity reversal. An analogy with a motion picture is again helpful: a film which showed the individual molecular motions in an expanding gas could be run backwards, and would then show these molecular motions as the gas appeared to contract. In either expansion or contraction the molecular motions would be in accordance with physical law. Hence, it is possible to postulate, as Loschmidt did, that the molecular velocities of some process for which H is decreasing may under other circumstances be reversed. In this latter process H would be increasing, in contradiction to the H-theorem. Both processes are physically possible, and hence the H-theorem cannot be valid, according to Loschmidt's argument.

Arguments which utilize the statistical nature of the H-theorem are required to explain the difficulty that Loschmidt finds with the theorem. We shall shortly see that the H-theorem is not to be taken as an inflexible statement that allows no exception, and that a state in which H is increasing with time is allowed.

The objection of Zermelo is instructive because an understanding and elucidation of the objection requires an emphasis on the statistical nature of the H-theorem. The argument is based on a theorem about dynamical systems which was proved by H. Poincaré for any *finite* system. The state of a mechanical system may be characterized by the values of the position variables, q_k, and the momentum variables, p_k, for the elements of the system. Certain values of the q_k's and p_k's are allowed by the energy of the system and by the constraints on the system. Poincaré's theorem requires that in sufficient time the system will approach as closely as is desired to any initial set of q_k, p_k values. Or, if the initial q_k and p_k values define some state S_0 of the system, the theorem assures us that at some later time the system will return to a state S that is as similar as desired to the state S_0.

Suppose then that a system is in a state S_1, with $H = H_1$, and at a later time comes to a state S_2, with $H = H_2$, where $H_2 < H_1$, in accordance with the *H*-theorem. The Poincaré theorem gives assurance that at some still later time the system will again be in a state that is, within some specified limits, the same as state S_1. But H_1 is greater than H_2. Hence, in reaching S_1 again there must have been an interval of time during which H increased, in violation of the *H*-theorem. In brief, the eventual return of any system to a given initial situation will involve a violation of the *H*-theorem. An example is provided again by our relatively simple system, an expanding gas. If the gas is initially concentrated in one region of a chamber, its subsequent behavior is to expand and fill the chamber. Poincaré's theorem requires, however, that in time the gas molecules will again be concentrated in the original partial region of the chamber. Such a "spontaneous" variation from uniform density is, of course, in violation of the *H*-theorem.

The *Wiederkehreinwand* brings out a clear contradiction between two theorems, the *H*-theorem and Poincaré's theorem, both of which are based on behavior properties of the constituent entities of a many-particle system. In the face of this contradiction the *H*-theorem must give way. The theorem is not to be taken as a rigorous, inflexible statement of what will occur, but rather as a statement of what will occur with high probability. The derivation of the *H*-theorem allows for such a relaxation from rigor, because, as we noted in section 1 of this chapter, the derivation rests on an assumption of randomness or chaos among the entities of the system to which the theorem is applied. Occasionally such randomness may by chance be significantly removed in favor of a state with some order among the entities, and a violation of the *H*-theorem may then occur. It is extremely unlikely, but it is possible, that the molecules of a gas may sometime all gather, momentarily, in one half of the gas container, in violation of the statement of the *H*-theorem.

Likewise, with a realization of the statistical nature of the *H*-theorem one can answer the *Umkehreinwand,* because we see that a reversal of velocities does not in itself involve a violation of the *H*-theorem. We must consider it to be a possibility that the molecules of a gas, for example, will have velocities such that entropy is decreasing from one instant to the next; for, the basic mechanical laws do allow any velocity or its reverse for a molecule, and the molecular chaos assumed for the *H*-theorem does not necessarily always exist. Just possibly, the streaming of a gas from a full to an empty cylinder may be reversed.

Boltzmann himself was very much concerned in his discussions of the *H*-theorem with the statistical nature of the theorem. The elements of a system, behaving in accordance with completely reversible equations, do generally combine to form irreversible processes, but the "generally" does allow for the exceptional and rare events, for which the system does not show irreversible behavior, with decreasing *H*. Thus Boltzmann wrote: "It can never be proved from the equations of motion alone, that the minimum function *H* must always decrease. It can only be deduced from the laws of probability, that if the initial state is not specially arranged for a certain purpose, but haphazard governs freely, the probability that *H* decreases is always greater than that it increases."[1]

The methods of statistical mechanics enable calculation to be made for the relative probability of sensible fluctuations from the generally observed course of behavior of many-particle systems. In a system with a large number of elements the probability of a change of the system for which *H* increases is actually very small—vanishingly small for practical purposes. Likewise, the time required for return of a system to some given original configuration of the elements of the system is extremely long if the number of elements is large. Boltzmann, for example, made the following calculation. A cubic centimeter of gas contains 10^{18} molecules, each of which, in a time *t*, is to return to an initial position and velocity, the average molecular speed of 5×10^4 cm./sec. remaining unchanged. Allowing the "return to initial condition" for each molecule to be within 10^{-7} cm. of each initial position variable and within 10^2 cm./sec. of each initial velocity component, Boltzmann estimated that the return to the original configuration would require $10^{10^{19}}$ years.[2]

The Poincaré theorem with the *Wiederkehreinwand* based on it, and also the *Umkehreinwand,* are important in showing the statistical nature of the *H*-theorem, and the possibility of exception to it, but we see that in ordinary application to systems containing many millions of entities the H-theorem expresses a virtual certainty. From a speculative point of view, however, with consideration of very long periods of time, or when fundamental scientific theory is being considered, the

[1]L. Boltzmann, Nature, *51,* 414 (1895).

[2]This calculation is presented in D. ter Haar's book, *Elements of Statistical Mechanics,* Rinehart and Co., New York, 1954, p. 342. Poincaré Recurrences are discussed in a paper of that title by Harry L. Frisch, *Phys. Rev., 104,* 1 (1956), for a simple one-dimensional system of elastic spheres. The recurrence time $\bar{t}$ is defined as the time required for a first recurrence of a given pattern of positions and velocities, within an error ϵ. He finds that $\bar{t}$ rises sharply as ϵ decreases, and that the fraction of time for which the law of entropy increase fails is proportional to a quantity $\geq \epsilon N$, where N is the number of spheres and $\epsilon < 1$. Hence, the greater the number of molecules "the greater can be our belief in the applicability of the laws of thermodynamics."

"virtual certainty" must be qualified with realization of the statistical aspect of the *H*-theorem. Nature everywhere presents processes that are to us irreversible, and the *H*-theorem makes it reasonable that they should so appear to us. Yet, these processes are—at least in many cases—made up of constituent reversible processes, and in principle the entire large-scale process is reversible, as specified by the Poincaré theorem. M. v. Smoluchowski suggested that we can therefore justifiably say of these processes that they appear to be reversible or irreversible according to whether or not a chosen initial state has a short or a long time of recurrence.[3] There is, though, a limitation too on the Poincaré theorem, a limitation that will be important in our discussion of time and the *H*-theorem in the next chapter; Poincaré's theorem is proven only for a finite system, and there is no assurance of eventual "return" for a system of infinite spatial extent.

3. ENTROPY AND THE QUANTITY H.

In Chapter II we saw that irreversible processes of a system are accompanied by an increase in a thermodynamic quantity of the system, the entropy S. The *H*-theorem asserts that the quantity H of a system will decrease, or at least remain constant, with the passage of time. The existence of general statements of increase and decrease respectively for those two quantities suggests that a relation exists between them. A detailed consideration of the role of H in statistical mechanical formulae, and of S in the equations of thermodynamics, shows that such is the case. The relation is in fact an exact one, and is given by the equation

$$S = -kH. \qquad (1)$$

In Eq. (1), k is the Boltzmann constant, 1.38×10^{-16} ergs/degree K, the same constant which gives the relation $E = \frac{3}{2}kT$ between the temperature T of a gas and the average external kinetic energy of a molecule of the gas. An increase in S may, then, be converted at once into a decrease in H, or *vice versa*.

The relation given by (1) further emphasizes the necessity of a many-element system for the *H*-theorem, and at the same time, gives greater extension to the entropy concept. We know that entropy, like all thermodynamic quantities, is by definition a characteristic variable for a complex system of many elements. Thermodynamics is a system of relations between variables of large-scale material systems; we cannot speak of the pressure, or temperature, or entropy, of a single molecule,

[3] See p. 15 of the paper by S. Chandrasekhar, *Rev. Mod. Phys., 15,* 1 (1943).

but only of these variables for a system containing many molecules, such that the effects of a single molecule are not discerned. The values of the thermodynamic variables are results of the many-molecular effects. The application of the concept entropy then implies that a complex system, made up of many elements, is under discussion. Therefore, Eq. (1), since it involves *S*, must refer to a many-element system. In reference to such a system, *H* does behave in general as required by the *H*-theorem, since the probability of deviation from the *H*-theorem is small when the number of elements of the system is large. In other words, Eq. (1) is valid for the kind of system to which it applies, and these are systems of many elements for which *S* may be defined and for which the *H*-theorem is true.[4]

But likewise, because of Eq. (1) we may apply the entropy variable to systems for which we do not have a strict thermodynamic definition of entropy. The thermodynamic variables of a system can usually only be measured when the system is in equilibrium. We can hardly, for example, assign a temperature to a system which is not in thermal equilibrium, since in that case different parts of the system have different temperatures. If the entropy of a system is determined in terms of thermodynamic variables, it is only when we know the temperature of the system, at least implicitly, that we can give a value to the entropy. But the quantity *H*, defined as $\int F \log F d\omega$, may be calculated for any state of a system, at equilibrium or otherwise, as long as an expression for *F* is available. Then, by means of (1) it is possible to go to a determination of *S* for the state. The statement that natural processes are accompanied by entropy increase is sometimes referred to as the "generalized law of entropy increase" when it is applied to processes that clearly involve non-equilibrium situations.

We have discussed individual particle equations with respect to their properties when their time variable is reversed, and also we have discussed reversible and irreversible properties in a context of thermodynamic concepts. We can now see some dependencies between individual-particle, statistical, and thermodynamic reversibility or irre-

[4]In sections 13-16 of *The Direction of Time* (see reference 6, of Chapter I), H. Reichenbach is concerned with the *Wiederkehreinwand* and points out that the fluctuations of a system will, over a very long period of time, give as many entropy increases as decreases; the entropy of the system cannot therefore be used for definition of time direction. In order to circumvent this objection, Reichenbach constructs his "hypothesis of branch systems;" he properly points out that in making time-entropy associations we are concerned not with the entropy of the total universe (which indeed may not even in principle be definable) but with the entropy of various quasi-isolated "branch systems" which we observe. In reaching his final definition of time direction in terms of entropy Reichenbach must assume, however, that the "vast majority" of the branch systems, during the period of observation, are moving from a low to a high entropy point. Further, the total universe is also assumed to be moving from a low to high entropy point during the period. His "branch hypothesis" is then in effect a statement of what is taken to be the common entropy-behavior of systems, against the possibility of an occasional random fluctuation which gives increase of *H* (or decrease of *S*).

versibility. A reversible thermodynamic process is an idealized case of observed natural processes, and it occurs in a maner such that the system undergoing the process is always in an equilibrium state. The entropy increase is zero for a reversible process; if the process is irreversible, there is an entropy increase. Now, shifting to a consideration of individual particle processes and the quantity H, we find that if the system is not in equilibrium H is then decreasing in time and the system is undergoing an irreversible process. That is, the effect of there being many particles is, as we have seen, to introduce an "irreversibility factor," even though the individual particle processes may be time-reversible. For irreversible processes, any decrease in H is parallel to increase in S.

The considerations which lead to the H-theorem also show that if a system is in equilibrium, then dH/dt must be zero, that is, H does not change. If the constituent, individual processes of the system are described by reversible equations, then at equilibrium the system would also be thermodynamically characterized as being at equilibrium and undergoing (thermodynamic) reversible processes, if any at all. The individual processes described by reversible equations may, therefore, combine to form either irreversible or reversible thermodynamic processes.

Individual-particle processes may also, we have seen, be described by time-irreversible equations. A many-element process made up of such elementary processes would almost certainly be a large-scale irreversible process, since one can hardly conceive of individual irreversibility leading to processes with large-scale thermodynamic reversibility. In summary then, individual particle irreversibility in a complex system would presumably always lead to thermodynamic irreversibility. Individual particle reversibility may give rise either to thermodynamic reversibility or irreversibility: in the former case there are equilibrium states and constant H and S; in the latter case, non-equilibrium states and changing H and S.

4. ENTROPY AND ORDER.

The equation relating H and S gives a specification of entropy of a system in terms which relate to the number of individual entities of the system in different velocity states, since H is defined in terms of the "distribution function" F. Another relation between entropy of a system was also obtained by Boltzmann; he showed that S and the thermodynamic "probability" W of the system are related by the equation

$$S=k \log W, \tag{1}$$

where k is again what we now call the Boltzmann constant.

The quantity W is defined as the number of possible arrangements of the elements of a system for some given thermodynamic state of the system, with any one arrangement as probable *apriori* as any other. Thus, for example, suppose a gas to be at thermodynamic equilibrium. The molecules of the gas, as regards the values of their position and velocity variables, could be shifted about in many, many different ways, without disturbing the large-scale thermodynamic equilibrium of the gas. That is, the number of arrangements W would be very large. If the gas were in a non-equilibrium state, such that its density, say, were D in half of the volume it occupied and $2D$ in the other half, then calculation would show that the number of molecular arrangements that give the state of prescribed non-uniform density is much smaller. In other words, W is much smaller in the case of non-uniform density, and hence, in accordance with (1), the entropy is smaller for the case of a non-uniform density. Left by itself the gas will, we know, go to the state of uniform density and larger entropy. Calculation shows that the number of possible molecular arrangements W is larger by far for a state of uniform density than for *any* state in which there are appreciable deviations from uniform density; this result agrees with the tendency of a gas to go to a state of uniform density, and with the principle that a system tends to go into its state of largest possible entropy, since S increases as W increases.

The creation of a sensible deviation from uniform density in a gas would require a partial ordering of the molecules from the state in which they give a uniform density. A freezing of a gas, such that the molecules took up definite, virtually motionless positions in the latticework arrangement of a solid crystal, would also involve a considerable ordering of the molecules as compared with their random motions in a gas. The crystalline material would have a much smaller W, and hence a much smaller entropy, than would the material in gaseous form. Here also, order is correlated with W. In general, a state for which only a relatively small number of molecular arrangements are possible is a highly ordered state, and a state with a large W is a less ordered state. Because of the relation between entropy and W we can, therefore, associate entropy with degree of order: as a system goes to states of higher entropy it is going to states of relatively lower order, or, of greater disorder.

The achievement of a relation between an originally thermal concept like entropy and the concept of order is impressive. With exact definitions of "arrangement" and hence of W—definitions which are fairly readily laid down—the relation becomes precise and quantitative. Qualitatively, one can appreciate the relation by considering typical physical processes. In the freezing of a gas, for example, order is clearly increased, as we have noted, and therefore entropy must decrease. This is in agreement with the thermal definition of entropy, as presented in Chapter II; if the entropy of a system is to be decreased, heat must be taken from it, and heat is removed from a system when it is frozen. Or again, if a gas expands by pushing against a piston it becomes less ordered, since the gas molecules are occupying a larger volume after the expansion than before. If there has been no temperature change in the gas, the entropy after expansion must therefore also be greater, since loss of order implies increase of entropy. But the temperature of a gas expanding against a piston can remain constant only if heat is supplied to the gas, since otherwise the work done by the gas in expanding cools the gas. The heat which must enter gives an entropy increase, calculated on the thermal definition, that is equivalent to the entropy increase that is calculated in terms of loss of order (or increase in W).

Using the entropy-order relation, one may say that the second law of thermodynamics asserts that natural processes always proceed with an increase of disorder. This statement of a fundamental natural law seems to appeal to non-scientists, in an age that has seen many evidences and threats of disorder in our society. We recall, however, that the second law requires an increase of entropy or disorder, only in a total isolated system; any one part may have a *decrease* in entropy at the expense of a greater increase in another part of the system. The "order" which is stored in fuel, in its chemically bound atoms, may be used to provide power which men utilize in building a highly ordered structure of building materials. Order then has increased among the building units, at the cost of order decrease among the entities constituting the fuel.

It is of interest to ask the speculative question, Is the universe as a whole tending toward a state of disorder? The second law of thermodynamics assures us that disorder is increasing as long as processes go on in the total universe. We need not therefore conclude, however, that the universe is necessarily moving without halt until an end state is reached in which all atoms are in a state of disorder comparable to those in a gas. There may well be great sources of order in the universe, sources which can give rise to an end state in which atoms are highly ordered, in terms of the standards we know for matter made up of atoms. We have learned of the atomic nuclei, for example, that they consist of elementary entities, with an array of order among them; that order is available to some extent for conversion into order elsewhere. It may be conceived that an effective equilibrium state be reached in which there are some elements of stable order in the universe, rather than a completely uniform disorder. Thus, in analogy, a chemically inactive atom is an ordered arrangement of particles, and yet is altogether stable, so long as its environment does not contain entities of sufficiently high energy to disrupt it. If the universe does now possess an over-all entropy that is relatively small, that is, with considerable element of order, then perhaps process may stop at a point where the total entropy yet remains small, with significant order present in the universe. This possibility will be discussed in greater detail in the final chapter of this book.

General References for Chapter IV.

R. C. Tolman, *Principles of Statistical Mechanics,* Oxford Univ. Press, 1938, Chaps. V, VI, XIII.

D. ter Haar, *Elements of Statistical Mechanics,* Rinehart, New York, 1954, Chaps. I, II, Appendix I.

P. and T. Ehrenfest, "Begriffliche Grundlagen der statistischen Auffassung in der Mechanik," *Encykl. d. Math. Wiss.,* Band IV, 4 Teilband, Art. 32, Teubner, Leipzig, 1907-1914.

S. Chandrasekhar, "Stochastic Problems in Physics and Astronomy," *Rev. Mod. Physics, 15,* 1 (1943).

Chapter V

Time and Our Universe

In the two preceding chapters we have discussed in some detail the reversible and the irreversible processes of natural phenomena. We shall now apply the general characterization of physical time which we presented in Chapters I and II; in the light of the further discussion that has been given of reversibility and irreversibility we will attempt some consideration of what the overall time aspect of our universe may be.

1. TIME AND THE H-THEOREM.

The *H*-theorem relates decrease in *H* to the time variable. Since we have emphasized that time is to be defined in terms of physical processes, and since the *H*-theorem states a temporal characteristic of a large group—perhaps the entire group—of irreversible physical processes, it might be objected that there is a circularity involved in the *H*-theorem. There is, however, no support for this argument. The time scale is defined by the cyclical processes which constitute our clocks and by the complex irreversible and progressively changing phenomena which exist in our natural world concomitantly with the cyclic processes. The *H*-theorem, applied to any given system of individual entities, makes use of a time variable that has been unambiguously determined without reference to the quantity *H*.

Because our time concept is determined independently of *H,* we are able to give physical meaning to possible fluctuations whereby the *H*-theorem is violated for a system. If a series of states of an isolated system occurred for which *H* increased rather than decreased with time, we would readily ascribe the increase in *H* to a violation of the *H*-theorem rather than to a reversal of time. Particularly, if the system considered were a small one, it would be a trivial statement that the *H*-increase represented nothing more than an unusual statistical fluctuation in the behavior of the system.

If the system considered were nothing less than the entire universe the independence of *H*-increase or decrease from the definition of time might be less obvious. However, in this case we appeal to the method for defining the direction of time which was presented in section 1 of Chapter II; later states of a system, in a series of states, are separated from a given present state by fewer states than are relatively earlier states of the system. This definition does not involve any consideration of *H*. For a series of states which does not include the present, the quantity *H* (like the entropy: see section 4, Chapter II) allows a determination of time direction among the states of the series by use of the *H*-theorem. Such a determination would of course rest on the assumption that the *H*-theorem was valid for the system—that is, that fluctuations which violate the theorem were not occurring.

2. RECURRENCE OF STATES.

The question of significant violations of the *H*-theorem is of interest when we consider the temporal nature of the universe. We recall that Poincaré's theorem requires that a dynamical system return to any given state if sufficient time elapses. Must we then consider that any given state of our world will in a long enough time occur again? A first comment is that the time spans involved would be far beyond even the time intervals of contemporary cosmogony. We saw in section 2 of the preceding chapter that the time required to bring about return of a given condition, in a system as relatively simple as a cubic centimeter of a molecular gas, was of the order of $10^{10^{19}}$ years. The estimated duration of our stellar universe as a universe with anything like the kind of matter we find today is only about 10^{10} years, a negligible figure compared with $10^{10^{19}}$ years. If we consider how much more complex our universe is than is a cubic centimeter of gas, and if we consider also that the universe is clearly far from an equilibrium state, the time that would be required for a "fluctuation" that would bring the universe back to some given state—say, as I find it to exist on June 20, 1957—becomes enormous beyond all time measures that we have experienced in the physical world. Thus, assuming the universe to be finite we can take the total number of atoms in the universe to be 10^{50} times the number in the one cubic centimeter of gas. For return to a given state we must wait until fluctuations occur such that in each of the 10^{50} groups of atoms there is a simultaneous return to that given state.

Still, in our attempt to understand what we can of the temporality of the universe we must consider every possibility which is not logically untenable. It is important to note that the Poincaré theorem is valid only for a system which is conservative, such that its total energy (and mass) is not changing, and also is finite, in that there are limits to the values allowed for the position and momentum variables of any particle. If the total universe is such a system, then we can expect that total states of the universe do recur. We then have a recurring universe of the kind envisaged by Giordano Bruno: every given state recurs again and again, presumably an infinite number of times. The enormous times required for each new occurrence of a given arrangement of the total universe becomes unimportant, as small or large durations, against the infinite temporal extension of the universe.

Even if the universe should fulfill the conditions of Poincaré's theorem, the continual recurrence of any given state would not imply that the universe had a cyclic behavior. Any given series of states, s_1, would presumably recur after a sufficient length of time. (There could be one or more states in s_1.) The states between two given occurrences of the series s_1 may be designated as a series of states s_2. In sufficient time the series $s_1+s_2+s_1$ would recur, and the states between the two occurrences may be designated as s_3. After states s_4 had occurred, the series $s_1+s_2+s_1+s_3+s_1+s_2+s_1$ would recur, and so on. The recurrence of any given series would always require a series of states not in the recurring series, since the system is not by definition cyclic with a finite period. We therefore will not arrive at a cyclic behavior of the system as a result of its random fluctuations, even though time goes on without limit. (We recall that in recurring cycles the same states occur in each cycle and the system returns to an initial state of the cycle at the end of each cycle—with no intervening states.) The recurrence of states in our hypothetical finite, energy-conservative universe does not therefore imply an atemporal cyclic universe; since states do not recur in a cyclic manner, there is an element of progressive change and hence a possibility for a progressive and unique forward time sense.

The preceding argument against the equivalence of eventual recurrence of states with completely cyclic behavior raises the following question: suppose a state *A* is followed by a series of states *b*, after which *A* recurs; will not series *b* again follow, since the system has the same arrangement of its elements in the second occurrence of *A* as in the first? It would then follow that any recurrence of a state implies that the system is a periodic one, with a fixed period between recurrences of a given state. The system would indeed be completely cyclical in its behavior.

This argument involves questions of specification of states. In the estimate made by Boltzmann for the length of time required for a given arrangement of molecules in a gas to recur, a "tolerance" of 10^{-7} cm. in position and 10^2 cm./sec. in velocity was allowed between the initial and the recurring state. If the tolerance had been made smaller, the estimated time for recurrence would be larger. The existence of some difference between two states, even though they are almost identical, means that succeeding states of the two may widely differ.

One might wish to go to the limiting case in which an initial state of a system and its recurrence are "exactly" alike. An expectation of such exact recurrence is not, however, physically tenable: for the position and velocity of the elements of a system cannot be specified with complete precision; or, more accurately, they do not, if current theory is correct, exist as quantities which can be specified with unlimited precision. There is, in other words, an element of necessary ambiguity in the specification of the state of a physical system, and hence recurring states do not have to give rise to similar series of consequent states and so give rise to completely cyclic behavior. This point of view, which clearly involves a divergence from a rigorously deterministic view, will be given further discussion in connection with our discussion of time and quantum theory in Chapter IX.

The eventual recurrence of states of the universe is perhaps not an appealing possibility—we would rather like to think that the universe is "going somewhere," and not merely recurring to past states in an apparently meaningless way. We will see in a later section that the weight of opinion today is hardly in favor of the recurrence notions; the assumptions of conservation of total mass and of a finite universe—assumptions which are necessary for the Poincaré theorem which ensures recurrence—can be strongly challenged. There is no certain argument, however, against the hypothesis of eventual recurrence of states.

3. TIME AND REVERSIBLE PROCESSES.

In Chapter III we discussed the possibility of time reversibility of physical processes. In section 1 of that chapter we considered the path reversibility property of an equation when we reversed the sign of the time *and* the velocity for a particle described by the equation. In that situation we carry a particle from point x_o and velocity v_o to point x_1 and velocity v_1, as time goes from 0 to t; in reverse, we go from x_1 and $-v_1$ to x_o and $-v_o$ as time goes "backwards" from t to 0. Such an invoking of genuine time reversal is somewhat unusual in physics; customarily, reversibility involves a reversing of velocity only, such that the particle starting at x_1 with $-v_1$ will reach x_o, with velocity $-v_o$, in an ordinary "forwards" time interval equal to the original interval from 0 to t. There seems however to be a possibility for a genuine reversal of time with respect to physically reversible processes with the result that we do reverse both time and velocity. We shall now discuss that possibility. (The achievement of path reversibility by *time* reversal only, without velocity reversal, involves contradictions, as was shown on p. 38 *et seq.*, and is therefore not to be considered.)

The progressively changing nature of our natural world, taken as the total system of which we have experience, defines a forward time sense. Likewise, any part or system of the world which we consider, and which undergoes an irreversible, progressively changing process, has a forward time sense associated with it. In contrast, a reversible process of a system or entity may not have the property of progressive change in terms of which a unique forward time sense can be defined. The reversible process may be timed, in terms of the time variable of the larger world about it, through interactions with that world. Thus, the freely falling particle is undergoing a reversible process. A man, who is part of the progressively changing world, watches the particle and determines its time of passage through various points; the light waves which pass from the particle to the man constitute an interaction between the particle and its surroundings. Observations of the starting and final positions of the particle are interactions which relate the process of the falling particle to the time variable of the surrounding world, and we have no hesitation in assigning times of starting and stopping for the particle.

If the reversible process is cyclically repeating itself, then within its own domain there is no forward (or backward) time sense from cycle to cycle; only if the process is making contact with the surrounding non-cyclical processes, can it be given a time sense. Thus, an electromagnetic wave going through a vacuum is a purely cyclical oscillation containing no intrinsic time direction. Likewise, as was pointed out in Chapter I, a system undergoing no change whatsoever contains within itself no time-defining process. (A state of not changing in any way may be taken as a degenerate case of undergoing completely cyclical change.) In particular, even a particle moving with constant velocity may be considered to be a system that does not contain any intrinsic definition of time sense, for it is only in terms of relations of the particle with its surroundings that the "time" of the particle may be defined.

There are special circumstances under which we may give a reversible process a time direction *that is forward or backward as we choose,* even though the process is making contact with surroundings in which a forward time sense is uniquely defined. Consider an electron of charge $-e$ to move from a point A to a point B, in a way such that a time Δt is required for the motion. If there is zero charge at A before the electron leaves at time t_0, then for $t>t_0$ the charge at A is $+e$, since charge $-e$ has been removed from A. Suppose that at B the charge is zero until the time $t_0+\Delta t$, when the electron comes to B and remains there. Then at B the charge is zero for $t<(t_0+\Delta t)$ and is $-e$ for $t>(t_0+\Delta t)$. (See Fig. 4a.) But we may also consider the process to be the motion of a positron from B to A, with time moving "backwards" from $(t_0+\Delta t)$ to t_0. (See Fig. 4b.) We assume, as before, that there is zero charge at A and at B for $t<t_0$. The positron leaves B at $(t_0+\Delta t)$

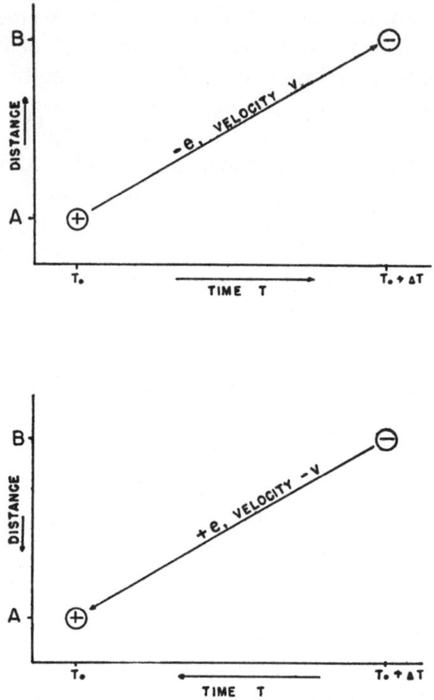

Fig. 4
4a: The electron leaves a (+) charge at A, at time t_0, and brings a (−) charge to B at time $(t_0+\Delta t)$. 4b: The positron leaves B at $(t_0+\Delta t)$, with the result that there is (−) charge at B for times after (t_0+t), and reaches A at t_0, giving a (+) charge at B for times after t_0.

and as time goes in a negative sense from $(t_o+\Delta t)$ to t_o the particle moves from B to A, reaching A at $t=t_o$. Then, in terms of the ordinary forward time scale, the charge at B is zero until $(t_o+\Delta t)$ and is $-e$ afterwards, because of the departure from B of the positron at time $(t_o+\Delta t)$. At A the arrival of the positron at time t_o brings a charge $+e$, and hence the charge at A is $+e$ for times $t>t_o$. With respect to observations that can be made we have then completely reproduced the situation of an electron moving from A to B by our backwards-in-time motion of a positron from B to A.[1]

[1]The conception of a positron as an electron travelling backwards in time was introduced into physics by Stückelberg and by J. A. Wheeler, and given further development by Feynman. See E. C. G. Stückelberg, *Helv. Phys. Acta, 14,* 588 (1941) and *15,* 23 (1942); R. P. Feynman, *Physical Rev., 74,* 939 (1948) and *76,* 749 (1949).

As the points *A* and *B* are arbitrary, we may say in general that the motion of an electron in free space between any two points is equivalent to the reverse motion, in reverse time, of a positron. "Equivalent" here means that physical observation will not detect any difference between the two motions. The positron motion is achieved by assuming a reversal of time with respect to the positron: its motion is taken to be described by an equation in which the usual sense of the time variable has been reversed; there is no reversal however of the time variable in the surrounding universe—that is, of the time variable of the physicist who is observing the electron (or positron) or of the laboratory clocks that he is using.

Several comments are in order on the hypothetical situation of physical time reversal which we have presented. First, we note that the reversal can occur only if we are considering motions which are described by reversible equations. Thus, if the motion of the electron from *A* to *B* were not in a vacuum but in a medium such that the electron lost kinetic energy as it passed from *A* to *B*, then the electron motion would be described by an irreversible equation; the reverse motion, from *B* to *A*, in which the electron gained energy from the medium as it moved through it would not be a physically plausible motion.

But motion which is path-reversible will not alone suffice to give observationally equivalent events when velocity and time are reversed. A light wave, for example, emitted at *A* at time t_0 and absorbed at *B* at time $(t_0+\Delta t)$, is completely reversible in its motion. The wave, however, carries energy from *A* at time t_0 and brings energy to *B* at time $(t_0+\Delta t)$; if there were a hypothetical reversal of velocity and time, the wave would carry energy from *B* at $(t_0+\Delta t)$ to *A* at t_0, and these energy changes could be observed to be in contradiction to the changes resulting from the *A*-to-*B* motion. In the electron-positron case the reversal in time is accompanied by a reversal in sign of the element (the charge) that is transferred, and physical equivalence of the two motions is thereby assured. That is, the bringing of an electron to a point is the same, with respect to charge at the point, as the carrying away of a positron: $+(-e)=-(+e)$, where "+" and "−" outside the parentheses indicate "bringing to" and "carrying away" respectively. If the electron transfer were being considered as the carrying of a particle of mass *m* from *A* to *B*, with no concern for the charge of the particle, then a physical time reversal would not be a plausible hypothesis, for the removal of mass at *A* and gain of mass at *B* would be observed effects that would not result from the motion with reversed time and velocity.

Genuine physical time reversal is a possibility, then, but only under the circumstances that the motion involved is reversible and that the time reversal may be counterbalanced by a reversal of sign in the quantity transferred. (Because of this latter sign change the reverse process is not the exact reverse of the original process.) The positron and electron satisfy these conditions. If there were particles of negative mass or negative energy, opposite to those of positive mass or energy, we would then be able to consider time reversal for chargeless particles.

The treatment of a freely moving positron as an electron moving backwards in time has been useful in physical theory, and there possibly is genuine physical significance to the time reversal that apparently converts a particle of a given charge into one of opposite charge. Also, it may be that the possibility of introducing negative matter or energy in association with time reversal will be found to be of significance.

The final and important point should be stressed, however, that we are here discussing time reversal only in the domain of reversible elementary processes. There is no inherent time direction associated with these processes, and therefore the time reversal of such a process is not a reversal of the time of the larger world surrounding the process. Thus, as we have seen, a free particle does not by virtue of its motion *in itself* undergo changes in terms of which a time direction may be defined; only in its interaction with the surrounding world may different times be assigned to the particle at different locations. It is therefore not the case when we speak of a time reversal, as in the electron-positron situation, that we are speaking of a reversal of the forward moving time of our entire physical world. We are speaking of the reversal of time only for a process which is time-reversible and not in itself a process that possesses a unique time direction.

The possible time reversal of a reversible physical process gains undue mystery and significance because of the use of the word "time" both with that succession of states of our world which defines our unique forward time sense and with the reversible time variable of reversible time processes. The reversal of the time variable of our total physical world is a conceivable possibility (as discussed in section 1 of chapter II) but any such reversal is sufficiently remote from physical observation or theory that in physics itself we do not consider the possibility.

The velocity- and time-reversal of a reversible process does not involve any reversal of the time-defining succession of different states of the world, and the reversible process does not in itself involve a time-defining process. It would then, perhaps, be well to use another symbol, say θ, for time in reversible processes. Theta-time would then not be the time of our physical world, but would be the time of those component reversible processes which are not defining a time sense. Of course, it is generally satisfactory to use the same symbol and concept for our forward moving time t as for individual reversible processes, since we generally assign a time sense to a reversible process through interaction of the process with the surrounding world. The existence of pairs of reversible physical processes which *allow* two possibilities for their time sense, even though the processes are in interaction with the surroundings, emphasizes that reversible processes, with their "theta-time," are different from those processes which can proceed in only one way as the physical world undergoes the state-to-state succession which defines the forward direction of our true time variable. And yet it must not be forgotten that a system of many elements, each of which is individually undergoing path-reversible processes, may present an overall irreversible t-defining process, as stated by the H-theorem.

4. THE PROBLEM OF THE AGE OF THE UNIVERSE.

The existence of a forward time progression in our universe means that there are physical processes of change which constitute the time progression. It is a natural question to ask, "Is there some epoch at which this forward progression of time began?" In terms of the physical processes that underlie our time concept this question of course becomes, "Is there a beginning point for the progressively changing physical process of our universe?"

If we are to find the beginning point, and the span of time from then until now, we must discern a progressively changing process which can be extrapolated backwards in time to its apparent beginning point. This process must be continuous, in that it persists as a distinctive process of the universe or some essential part of the universe. Until a few decades ago no such process was apparent. In the nineteenth century geologists began to understand successive stratifications of the earth's crust as a process which extended through much of the earth's lifetime, and they were thereby able to make estimates of the age of the earth. In the past few decades the estimates have been corroborated and refined by studies of the decay products of radioactive deposits in rocks. A process which will give an estimate of the age of the earth is of immense interest to us, but we are not thereby necessarily helped in making an age estimate for the entire universe. The terrestrial and stellar universe seems to consist of so many and varied complex processes and systems that one might reasonably say that there is little hope of finding a progressively changing process that is essential to the universe, such that consideration of the development associated with that process will give a temporal measure of the universe. And yet, remarkably enough, there actually is a process which perhaps does, at least in some partial sense, give us a measure of the age of the universe.

The process that we refer to is the observed motion of the various groups of heavenly galaxies away from our own local group of galaxies. This motion is, as far as observations have gone, a universal aspect of our universe. It is, however, an aspect that is apparent only when the greatest distances known to astronomy are involved. Our own Milky Way galaxy has a diameter of about sixty thousand light years, and it is one of a "local group" of galaxies, the other members of which lie within a distance of a million light years or so from our Milky Way galaxy. It is between our local group and other galaxies or groups of galaxies that the recessional motion is observed. The spectra of light from far galaxies show a "red shift"—that is, a displacement of spectral lines toward the red (or long wave-length) end of the spectrum. There is no other reasonable interpretation of this displacement than that it results from a motion away from our own galaxy; for, in accordance with the well-established Doppler equation for electromagnetic radiation, light which is coming from a source that is moving away from an observer is seen as having been shifted to longer wave-lengths, in comparison with light coming from a similar source at rest with respect to the source.

The motion of the far galaxies is found to be closely proportional to the distance of the galaxy from the earth: the greater the distance the greater the red shift, and, hence, the greater the recessional motion. The velocities of recession become very large for distant galaxies: for a nebula in Virgo, about 30,000,000 light years distant from our galaxy, the velocity is some 1200 kilometers per second. For a distant galaxy in Hydra, 1,300,000,000 light years away, the velocity is about 60,000 km/sec, and is therefore approaching the limiting velocity for matter, namely, the velocity of light, 300,000 km/sec. The proportionality constant, which when multiplied by the distance gives the recessional velocity, may be calculated from the two quoted velocities to be about 5×10^{-18} sec^{-1}. This factor has been found to be substantially a constant for all those far galactic bodies for which distance and recessional velocity have been determined.

It should not be thought that our Milky Way galaxy is shown by the galactic red-shift to be in the center of the universe, with galaxies moving away from it with velocities proportional to their distance. Mathematical analysis of the motions of points in a volume which is expanding uniformly shows that any point in the volume will be moving away from any other point with a velocity proportional to the distance between the two points.

The discerning of the uniform expansion of the galaxies is perhaps the outstanding empirical discovery of our age. Pioneer work on the galactic motions was carried out by V. M. Slipher; E. P. Hubble and M. L. Humason clearly established the universality of the red-shift. The proportionality constant between distance and velocity, evaluated as about 5×10^{-18} sec^{-1}, has come to be known as Hubble's constant, H. The value of its reciprocal is about 6×10^{9} years, and, if we assume that recessional velocities have remained the same in times past, multiplication of that number by the observed velocity of a galaxy gives the distance of the galaxy from the earth. That is, whereas $H \times$ Distance = Velocity, Velocity $\times\ T =$ Distance, where T is $1/H$.

It is easy to see that the recessional motions of the galaxies suggest an extrapolation back to a hypothetical beginning point at which all of the vast galaxies of the universe were condensed into a relatively small volume. Expansion may be thought to have begun from this point, in an event akin to a great explosion, with matter moving outwards with varying velocities; the fastest moving masses are now farthest from us, and any one of them is at a distance given by its velocity multiplied by the reciprocal of Hubble's constant, $T = 1/H$. That quantity is clearly then the time that has elapsed since the event with which outward motion began. It is tempting to call T the "age of the universe." Studies of the age of the earth indicate a value of about 5×10^{9} years, and this result gives confirmation to the age interpretation of T, since it might reasonably be thought that the earth and other planets would be formed early in the history of our galaxy; at the very least, the galaxy must be no younger than the earth.

We do not wish here to enter into a detailed discussion of the problem of the age of the universe, except to give some indication of the possibilities that have been suggested on the basis of astronomical observation and physical theory, for the time characterization of the cosmos. The mutual expansion of the galaxies is clearly a key fact in any proposal of a cosmological model; and yet, it is also clear that the time T that is gained by a simple extrapolation of the expansion process will not pass as the "age of the universe" without considerable criticism.

Even assuming that we are justified in saying that T years ago the universe began in a highly condensed state, we cannot of course properly call T the age of the universe, for presumably that condensed state itself had a past. That is, we cannot so consider T unless we wish arbitrarily to say that there was an event of total creation which gave birth to the universe T years ago. Some cosmological theorists have taken this point of view, but with little scientific justification. We have learned in science that the consideration of natural events as resulting from preceding events—the conditions of succession being set forth in natural law—is in general justified by observation of nature. The postulation of a creation event, which arose without prior conditions and which gives no possibility for scientific investigation of its preceding states, is not in keeping with scientific practice; although to be sure, if there came to be compelling evidence in favor of such a creation event, then we would accept it, for the procedure of science is never inflexible.

We might much better refer to T as the "age of expansion" in our universe, rather than as the age of the universe. There should be criticism, however, even of this more cautious ascription to T. It is possible that expansion began not at a highly condensed state but at some state intermediate between maximum density and the present distribution of galaxies. For example, the universe of galaxies may have been in a state of relative equilibrium when the over-all galactic density was one-half of its present value, and expansion may have started from this state rather than from one of extreme condensation. Such an hypothesis implies, of course, some factor that would start the uniform expansion which we now observe. In general, observationally obtained values of H, even with the addition of some possible information on changes in recessional velocity in time (obtained from observations of distant galaxies) is not sufficient to give us a value that may justifiably be called the "age of expansion." A physical model or theory of the universe, which tells us what the past history of the expansion has been, is also required.[2] The simple extrapolation, then, that gives the "age of expansion" as $1/H$ is only for an idealized, limiting case that most probably is far from representing the actual past motions of the universe of galaxies.

The general relativity equations for the relations between matter and the overall curvature of cosmic space have solutions which require that the radius of curvature of space increases in time; cosmological models in agreement with the observed expansion of the galaxies may therefore be based on the relativity field equations. Given suitable initial conditions for the expansion, it is also possible to obtain solutions of the field equations which require that the radius of curvature oscillates between a minimum value and some very large value. In other words, the theory of general relativity

[2]See "The Second-Order Term in the Redshift-Magnitude Relation," by Fred Hoyle and Allan Sandage, *Publ. Astron. Soc. of the Pacific, 68,* 301 (1956); also "The Theoretical Aspects of the Nebular Redshift," by H. P. Robertson, *ibid., 67,* 82 (1955).

allows the possibility that the expansion of the universe of galaxies which we observe will eventually come to a halt, and will be followed by a period of contraction, these alternate phases of expansion and contraction continuing indefinitely.

Such an oscillation-model of the cosmos requires as one of its assumptions that the universe is finite in radius. Astronomical observations do not give any clear indication counter to this assumption, and they do rule out the possibility of a universe which extends infinitely and everywhere contains a star density comparable to that which we find in our observable universe. Calculation shows that if stars which deliver light to us were distributed everywhere in an infinite space we would then find the light of the sky to be as bright as the surface of a star.[3] Only if the effective star density fell off at least as fast as the inverse first power of the distance from our galaxy could we have an infinite universe of stars. The density decrease could be either a literal decrease in stellar matter, or a result of recessional velocities that approached the speed of light for the very distant galaxies, with accompanying Doppler-shift decrease in radiant energy received from those galaxies.

The general theory of relativity, although incomplete, may be regarded as a valid description of the relation between space and matter. The various kinds of cosmological models which are derived from the general-relativity field equations are therefore not to be dismissed as idle speculation. At present, however, there is little empirical evidence for deciding between the various possible models: we cannot say whether the correct model is one in which expansion continues until the galaxies are indefinitely far apart, or is one in which there is oscillation between a maximum and a minimum radius; nor can we say whether the total radius of the universe is finite or infinite.[4]

A possibility for the age of the universe in addition to those presented by general relativity theory has been introduced in the so-called "steady-state" cosmological theory of F. Hoyle, H. Bondi, and T. Gold.[5] This theory is not inconsistent with general relativity, but

[3]This is "Olber's Law." See H. Bondi (Reference 3, Chap. I), Chapter III.

[4]Although observational results for several decades have consistently confirmed the existence of a red-shift proportional to distance for galaxies, astronomers have in the past decade found reasons, based on observation, for systematically increasing their galactic distance values by roughly a factor of 2. The values of H (and T) which we have used in this book are on the basis of these new distance estimates. Allan Sandage has reported new observational studies which lead him to suggest that the value of H should be yet further decreased, so that $T = H^{-1}$ is in the neighborhood of 10 rather than 6 billion years. Present achievements and problems in cosmology are surveyed in R. Schlegel, *Completeness in Science*, Appleton-Century-Crofts, New York, 1967, Chaps. 7 and 8.

[5]H. Bondi and T. Gold, *Monthly Notices Roy. Astron. Soc.*, *108*, 252 (1948); F. Hoyle, *ibid.*, *108*, 372 (1948). An early form of the theory, developed by W. D. MacMillan, is discussed by R. Schlegel, *Amer. Journ. Physics*, *26*, 601 (1958).

does assume a new process that is not described by the equations of standard relativity theory, the process of spontaneous creation of matter in empty space. According to the theory creation occurs at the rate of about one hydrogen atom per liter of space in every 10^9 years. This rate of production, although exceedingly small from our laboratory point of view, produces sufficient atoms to form galaxies in the number density that we observe. There are of course uncertainties in the account astronomers give of the development of stars and galactic systems, but a plausible story can be made for the condensation of atomic gas into stellar systems. The expansion of space is an associated aspect of the steady-state theory; the heavens do not become increasingly filled with stars, because the galaxies stream away from each other, as observationally indicated by the galactic red-shifts.

The essential point of the steady-state theory is that matter is continually coming into our observable universe through the postulated spontaneous creation process, and is continually leaving through the recession of galaxies from any given observer, the galaxies eventually passing "over the horizon" and out of possible view to the observer. All parts of the universe are, then, in a steady state with respect to such large-scale features as average density of galaxies. Observers at any place in the universe, at any time, would see the same average galactic features. For any given observer the universe is finite, since the galaxies beyond a certain distance have receded out of the observer's view, but conceptually the entire universe is spatially infinite. The universe extends without limit in time; and since in its large scale features it is always the same to any observer, the universe lacks progressive change in its large-scale, galactic properties.

The steady-state theory has seemed to many people shockingly to disregard one of the best established tenets of physical science, the law of conservation of mass-energy. On the other hand, the theory does have the advantage that in principle it brings creation of matter into the domain of scientific investigation; and it is perhaps not far-fetched to assume that space, which does propagate electromagnetic waves that can constitute themselves into positrons and electrons or other forms of matter, may also on relatively rare occasions spontaneously give rise to matter. In any event, the steady-state theory is subject to empirical test, in that the theory has implications for the spatial distribution of stars in different stages of development, for the formation of the elements in the stars (the formation must be a continuing, contemporary process, according to the theory), and for the space distribution of galaxies, which must on the average be uniform throughout our observable universe.

The properties of the recently discovered quasars are still a matter of controversy; but by one interpretation of the observations the quasars are stellar objects at remote distances and times, unlike anything found in the regions of the universe nearer to us and hence in violation of the large-scale uniformity that is required by the steady-state theory. A further empirical fact, counter to the steady-state theory, is the discovery in 1965 of a cosmic background black-body radiation which quite plausibly is residual from an early, condensed state of the universe.

5. OVERALL TIME PROPERTIES OF THE UNIVERSE.

In Chapter II we set forth the various possibilities for the temporal characteristics of systems in which various kinds of processes are occurring, and in the chapters following that one we have examined some of the time-defining features of the natural world. We now wish to consider the universe as a total system, and by taking account of the kinds of processes that may be presumed to occur in the universe we will attempt to make some estimate of its *possible* overall time characteristics.

It is obvious that processes of progressive change exist in the domains of nature that we have been able to observe; and likewise, it is apparent that there are cyclical processes in our natural world. The first judgment to be made of the temporal characteristics of the universe is therefore that the universe by virtue of its progressively changing processes is intrinsically temporal—that is, time extends infinitely in past and future.

The *H*-theorem suggests an obvious variation of the preceding time characterization; in the words of Boltzmann, the *H*-theorem would imply "that the universe is generally in a state of equilibrium, and regions of non-equilibrium, like the region of the universe known to man, are very improbable; but with the universe sufficiently large the probability of there being such regions increases." [6] Any region, then, which is not at equilibrium—as clearly is not the observed world about us—would be expected in time to reach a state of equilibrium: a state of absence of change and hence absence of time. This eventual state of equilibrium may likewise be predicted for the universe by entropy considerations. Since the entropy of a non-equilibrium system is always on the increase, we should expect that the universe will eventually reach a state of maximum entropy; in this state all heat or energy flow will have ceased because all energy differences between different parts of the universe will have been erased. The universe will "nowhere be white or black, but will everywhere be uniform gray." This is the idea of the "heat death" of the universe, a concept which was much discussed in the period around the beginning of this century.

The use of the *H*-theorem, or of the entropy-increase principle, as justification for the prediction of a final equilibrium state does

[6] L. Boltzmann, *Nature, 51,* 414 (1895). Boltzmann here writes that this is "an idea of my old assistant, Dr. Schuetz."

require the further assumption that we are considering a "closed" universe in which new elements are not entering the universe and preventing the achievement of equilibrium. The spontaneous formation of new matter-energy as in the steady-state universe, or the initiation of new interactions with non-equilibrium regions, in a totally infinite universe, might be such disturbing elements. Further, it is a corollary of the concept of a progression toward equilibrium that the universe must have been at a state of low entropy, and therefore high degree of organization, at some time in the distant past. Associated with this concept, we have the problem [7] of understanding how the universe came to be in the state of low entropy from which it began that process of changing toward higher entropy which we find to be characteristic of our observable universe.

We see, then, that the consideration of entropy increase, applied to a universe that is "closed" in the manner specified above, does put a limitation on the infinite time extension that is suggested by a simple extrapolation of the observed progressive change in our universe; the universe begins with some highly organized state of low entropy and will come to an equilibrium state of maximum entropy, after which change and therefore time will cease. However, even with the assumptions that have been made, these limitations cannot be set down without reservations; for, if instead of using the entropy principle we go to the H-theorem in which the *statistical* aspect of entropy increase or H-decrease is not neglected, we see that there is a possibility of continual change even in a closed universe. The entropy principle, as a theorem of thermodynamics, is a generalization from the observation of large scale processes and does not allow any possibility of processes that violate the principle. On the other hand, we have seen that the H-theorem does permit a possibility, however slight, of increase rather than decrease of the H-function of a system for successively later states of the system.

The H-theorem is derived for a system in which the basic processes are path-reversible. The basic processes of our universe may all be characterized by such reversibility, although we saw in Chapter III that it is by no means a necessary characteristic even of classical mechanics that all processes described by it are path-reversible. But if we assume path reversibility, and also assume that the universe is finite and that the conservation principles hold for it, then the Poincaré theorem will require that fluctuations of the system of the universe will in time bring the system arbitrarily close to any previous state. The enormousness of the times that would be required for recurrence of a state has been discussed in section 2 of this chapter.

[7] A. S. Eddington, *Nature of the Physical World,* Macmillan, New York, 1929, pp. 83-84.

Also, we saw in that section that recurrence of states does not imply cyclical repetition of states, and hence the recurrence would not destroy a progressive (acyclical) change of states with its continuance of time in the universe. In a word, for a universe whose basic physical processes are path-reversible processes, fluctuations among the atoms or other entities of the universe would not allow equilibrium to persist; there would always be a tendency toward equilibrium, but such equilibrium once reached, would be destroyed by fluctuations which for a time increased rather than decreased *H*. We therefore would not expect a "heat death" in an equilibrium state of the universe. And, in the spirit of the quotation from Boltzmann, any subdomain of the universe which reached an equilibrium state would eventually by fluctuations reach any specified non-equilibrium state.

As a summary of our discussion thus far in this section we can say that the progressive change which we see in the universe suggests extrapolation of that change, and a corresponding passage of time, to infinite past and future. Thermodynamic considerations suggest, however, that there are limitations in the past and future on this time extension, particularly if the universe is finite and is not losing or gaining mass and energy. But, if the universe is finite, and conservative of mass-energy, and also is constituted by elementary processes which are path-reversible, then the statistical considerations of the *H*-theorem suggest that the thermodynamic limitations on time extension are not absolute but only indicative of a tendency, which would in fact in sufficient time always be counter-established.

There is a kind of emptiness in these discussions of possible time behavior of the total universe, because we clearly are going well beyond matters about which we have adequate observational data for a decision between various possibilities. And yet, there is value in setting out the possibilities which have any plausibility at all, so that we may better see what is the import, for time characterization, of the empirical information that we do have.

It may be that the universe is not a "closed system," and that applications of the entropy principle or the *H*-theorem to the universe, considered as such a system, are altogether wrong. There may be a basic process in the universe which is fundamentally changing the system of the universe, and thereby giving a basis for a continuous, forward time direction, with no indication of eventual equilibrium or recurrence. The expansion of the universe might be such a fundamental progressive change. The entropy principle is certainly not then applicable, for the universe is then not a closed system; for one thing, the increasing energy of the galaxies as they move away from each other violates the ordinary conservation-of-energy principle. If the expansion is to continue indefinitely, perhaps with all galaxies eventually receding from each other with the speed of light, we then have surety for an indefinitely continuing forward time sense in the continuous expansion of the galaxies.

The steady-state theory of the universe provides a continuous process, in the formation of new matter, with which a continuous forward time sense may be associated. The amount of matter is increasing steadily, and the progression of time could be associated with the amount of matter. There are difficulties with such association, in the steady-state theory as it is at present enunciated. Each observer, according to the theory, always sees approximately the same total amount of matter in his total observable universe and yet, in the total conceptual universe the amount of matter is increasing continuously. There are here difficulties of divergence of the total matter or space in the entire conceivable universe; each cubic centimeter of space must be thought of as becoming in time the space and matter of an observable universe, and each cubic centimeter of that universe could in time itself be an observable universe, and so on! On the other hand, the *observable* universe in steady-state theory is not changing, and perhaps properly modified H-theorem-entropy-principle considerations could be applied to this observable universe, with some consequent modification for that observable universe of the extrapolation of progressive change into an infinite future.

We have been considering the universe as containing—at least in some degree—both non-cyclical (for defining time direction) processes and cyclical processes (for quantifying time). The universe, so considered, has the temporal characteristics of a system with both cyclical and non-cyclical non-reversible processes, as given in the summary Table of section 6, Chapter II. We shall now consider some of the other possibilities for the system of the universe as listed in that Table. But before doing so it will be well to recall several points about reversibility and periodicity.

We know that non-reversible, entropy-increasing processes occur in the universe, and further, we know from the H-theorem that even elementary processes which are truly reversible will combine to give rise to non-reversible processes. Also we note that reversible processes are not to be confused with cyclical processes. A reversible large-scale process, involving many constituent particles, is by definition without entropy increase or decrease. Such processes may not actually occur in nature (see discussion in Chap. II, p. 24), but in any event they need not be cyclical; one can, for example, at least conceive of a gas expansion which proceeds reversibly—with no "internal friction"—and is in no way a cyclical process. Likewise, cyclical processes need not necessarily be reversible ones; a cyclical, non-reversible process may be maintained at the expense, as it were, of energy exchange with the surroundings, that is, at the expense of entropy increase in the system including the cyclical process and its surroundings. We saw, however (Chapter II, p. 30), that a system cannot maintain only cyclical, non-reversible processes.

The table of Chapter II indicates three possibilities for systems whose processes are reversible, with non-reversible processes assumed to be excluded. Since, however, we do find non-reversible processes in nature, we for the time exclude those possibilities for the temporal characterization of the universe, although we shall presently give them some consideration.

For a system with non-reversible processes, the Table also lists three apparent possibilities. The obvious possibility—that the universe contains both cyclical and non-cyclical (progressively changing) processes—has been considered in detail. The existence of a system with non-reversible processes that are all cyclical is not actually a possibility, as we have seen. Hence, there remains only the possibility of a universe containing non-reversible processes, all of which are non-cyclical. This universe would contain acyclical processes which uniquely define a time direction, but would not possess uniformly periodic processes which could be used as a basis for a time scale. Our observations have generally indicated that the universe does contain a consistent set of uniform cyclical processes, on which we may base a true time scale. However, as we have discussed in section 3 of Chapter I, it is possible that over a very long period of time there may be changes in the relative frequencies of various kinds of processes. At the time of this writing it is hoped, for example, that with the extremely accurate cesium clock it will be possible to determine whether or not there is any change in gravitational clock frequencies with respect to the frequencies of electromagnetic (atomic) clocks over the course of a year.[8] If it should be found that the various fundamental cyclic processes of our universe do not bear a consistent relation to each other, we will arbitrarily have to choose one type of process as defining the time unit. We will then have the situation that there is no unchanging cyclical process which defines a time scale in the universe, unless—and this is not improbable —physical theory shows that one set of cyclical processes is fundamental to other sets of cyclical processes, in a manner such that the fundamental set determines the properties of the others.

We now consider the universe as a system whose processes are reversible. We excluded the temporal characteristics associated with reversible processes, on the grounds that non-reversible processes are clearly present in nature. The existence of expansion of the

[8]See the discussion by E. C. Bullard, *Nature, 176,* 280 (1955). Also, see D. H. Wilkinson, "Do the 'Constants of Nature' Change with Time?," *Phil. Mag., 3,* 582 (1958).

universe suggests, however, that on a cosmological scale—a scale on which the individual units of the system are galaxies—the system of the universe may be undergoing only a reversible non-entropy-producing process. This process would be the motion of the galaxies away from each other, or the expansion of the universe. This process might be non-cyclical, as say a uniform expansion; in this event the reversible cosmological process would still provide a process of progressive change for definition of time direction, perhaps in addition to non-cyclical, irreversible processes occurring on a smaller scale. We also, however, have the possibility that the reversible cosmological process is cyclical; we saw in the previous section, for example, that a universe in which there are alternating cosmological expansions and contractions is described by one solution of the general relativity field equations. On a *cosmological* scale such a universe would be atemporal, since there would be no progressive change from one cycle of contraction-expansion to the next. Within any one cycle there would indeed be the progressive change associated with the cosmological process, as well as with the irreversible processes that occur in the universe, and hence there would be a proper time direction within any one cycle. But if the cosmological process were truly cyclical, with no thread of acyclical process continuing from one cycle to the next, we would be obliged to say that the universe in its overall aspects is not characterized by a past-to-future time sense.

The description "atemporal" does seem to be correct for subdomains of the natural world in which the physical processes are purely cyclical. The hypothetical universe which repeats some grand cycle is likewise properly atemporal, even though our familiarity with a forward moving time concept, based on the progressively changing natural world that we know, tends to make us intuitively feel that each cosmological cycle is "later" or "earlier" in time than other cycles. The *observable* universe of steady-state theory might also be characterized as atemporal, since in its cosmological characteristics it does not undergo any progressive change. The ascription of atemporality is less plausible, however, than in the case of the periodically expanding and contracting universe, since it seems likely that there would be a continuous development of an irreversible, acyclical process from epoch to epoch of the observed steady-state universe. It might be argued, however, that the history of stars and galaxies justifies a view that our own Milky Way galaxy is going through an evolutionary process similar to that of all other galaxies, and that the earth, as well as any other planet, will eventually be dissipated in some final dispersal of the galaxy. The time extension associated with the physical aperiodic processes of our earth and galaxy would therefore be limited to the extent of those processes, with no continuation of system or process from one galactic history to the next. The observed steady-state universe would eventually end, and then occur again, but always with each new occurrence altogether similar on a cosmological scale to previous occurrences. In the atemporal universe of expansion-contraction cycles it would presumably be just those expansion-contraction processes that would bring about an end to all galactic development and likewise again bring conditions in which new galaxies would be formed.

We thus conclude our consideration of the temporal characteristics of the system of the universe. Any judgment or choice with respect to the various possibilities is highly speculative. It is however to be emphasized that the achievement of further empirical information about the physical world should bring further evidence for or against the various possibilities. The discerning of the temporal nature of our universe rests ultimately on our gaining knowledge of the kind of physical processes that occur in the universe; philosophical discussion of time may give clarification but cannot give a solution to the problem of time and the universe.

We have now considered the concept of physical time, using as the basis of our consideration the dependence of the time concept on the existence of cyclical and acyclical processes. Further discussion of the time concept and its relation to the physical world has been based on the distinction between reversible and irreversible processes, the entropy principle, the *H*-theorem, and physical information about the kinds of processes that exist in nature. We are, however, by no means finished with our task. Both of the two great theoretical developments of twentieth century physics, the relativity theory and the quantum theory, have relevance for the problem of physical time. Relativity theory especially has had fundamental effects on our concept of time, and we shall consider those effects in the next three chapters. We shall see that we learn something quite new about time in relativity theory, although, also, we will argue that the implications of the theory do not essentially disturb the discussion of time that we have so far presented.

General References for Chapter V.

H. Bondi's *Cosmology,* referred to in the text, gives a good introduction to the problems and achievements of cosmology. A detailed discussion of cosmology and relativity theory is presented in R. C. Tolman's *Relativity, Thermodynamics, and Cosmology,* Oxford Univ. Press, 1934. See also the review paper, "Relativistic Cosmology," by H. P. Robertson, *Rev. Mod. Physics, 5,* 62 (1933).

"Age of the Universe" issue, *Brit. Journ. for the Philos. of Science,* V, No. 19 (Nov., 1954).

Henry Margenau, "Can Time Flow Backwards?," *Philos. of Science, 21,* 79 (1954).

M. and G. Burbridge, *Quasi-Stellar Objects,* W. H. Freeman, San Francisco, 1967.

Chapter VI
Relative Motion and Physical Processes

Effects of relative motions between systems are considered in the special, or restricted, theory of relativity. It is in this first development of relativity theory that Einstein presented a concept of time which differs markedly from the time concept of Newtonian physics, and it is the special relativity theory which has become an integral part of the theory and practice of physics. The evidence for the theory, as it is used in physics, is overwhelmingly in its favor.

The general theory of relativity appears to be an altogether correct theory of gravitational forces, and the theory is clearly of great significance in the development of a complete theory of physical space-time; but the general theory is incomplete, and so far it has not yielded much of import for our concept of physical time beyond what is contained in the special theory.

We shall present in this chapter a discussion of the elements of the special theory, and an account of physical observations which confirm the "time transformation" of the theory. The next chapter will be concerned with some paradoxes of the relativistic time concept —particularly with the so-called "clock paradox" that has persistently dogged the heels of the theory. In Chapter VIII we shall discuss the validity of the extrapolations of the relativistic time transformation to large-scale clocks and other complex systems. This discussion will rest in part on a careful consideration of the laboratory confirmations of the time transformations, and the descriptions given of these experiments in the present chapter are therefore of importance for the arguments of Chapter VIII.

1. THE SPECIAL THEORY OF RELATIVITY.

Relativity theory had its origin in the problem of what would be the electrical and magnetic effects resulting from motion through space at high speeds. At the time of the birth of the theory it was assumed that space was pervaded by the "ether," a tenuous material which had the property of forming electrical and magnetic fields and of supporting light and other electromagnetic wave vibrations. It could be assumed that bodies move with respect to the ether, taken to be a kind of placid, all-pervading sea. Experiment, however, failed to detect any effects of motion through the ether, even though in theory the effects should have been observable. A further possible assumption was that motion of a body through the ether is not detectable because the motion itself causes the ether to change the body in such a way that the effects of the motion cannot be observed. In 1903 H. A. Lorentz, in what are now the famous "Lorentz transformation equations," found precisely what must be the effect of the ether on the space and time variables of the moving body, if no effects of the motion are to be detected.

H. Poincaré and A. Einstein, in separate papers published in 1904 and 1905 respectively,[1] took the step of suggesting that electrical-magnetic properties of a system are unaffected by uniform motion of the system, and that the ether hypothesis may be dispensed with; and further, that the changes described by the Lorentz transformations are physical changes arising from relative motion between two systems. Both Poincaré and Einstein pointed out the major implications of accepting the Lorentz transformations for physical properties of relatively moving objects. Einstein gave a notably complete discussion in his paper, including an analysis of a new possibility for our concept of physical time, and he therefore is properly regarded as the major author of the special relativity theory,—though, indeed, Einstein "stood on the shoulders of giants," to use the words Isaac Newton had applied to himself some two hundred years earlier.

The preceding is a very brief discussion of the historical development of the special relativity theory. We can give a logical rather than historical presentation of the theory by discussing it as a consequence of two assumptions: (1) the state of uniform motion of an inertial system has no effect, for observers at rest in the system, on bodies that are at rest in the system; (2) the velocity in a vacuum of light or other electromagnetic radiation from a given source is the same for observers in any inertial system, regardless of any uniform motion of an observer's system with respect to the source. By an "inertial system" we mean one which is free in space, without subjection to physical forces of any kind (as gravitational, electromagnetic, or nuclear).

The first of the above two assumptions is not a surprising one. It

[1]H. Poincaré, *Bull. des Sc. Math.,*[(2)] *xxviii* (1904), p. 302; English translation by G. B. Halsted, *The Monist,* Jan., 1905, reprinted in *The Scientific Monthly, 82,* 165 (April, 1956). A. Einstein, *Ann. der Physik, 17,* 891 (1905); English translation by W. Perrett and G. B. Jeffery in *The Principle of Relativity,* Dover, New York. The historical development of the special theory of relativity is described in detail (although with a surprising lack of emphasis on the work of Einstein) by E. T. Whittaker, *History of the Theories of Aether and Electricity,* vol. 2, Chapter II, Nelson, Edinburgh, 1953.

tells us, for example, that physicists who are making observations and doing experiments on physical events in their systems will get the same results whether their laboratory is at rest with respect to the earth or is moving uniformly in a railroad car. Another way of putting the assumption is to emphasize that there are no "signposts in space" which tell an observer in any way whether he is moving or at rest, as far as uniform (non-accelerated) motion is concerned. It is only by referring to some reference system, such as the earth or the sun, that an observer can determine what is his state of motion or rest. In a word, uniform motion is relative, not absolute.

The second assumption is a much more drastic one. Suppose that observer S is at rest with respect to a light source A, and observer S' is moving away from the source. Common sense seems to tell us that if S and S' both make measurement of the velocity of light coming from A, S' will find a smaller velocity than will S. The assumption denies this, and in fact asserts that S and S' will both measure exactly the same velocity. The validity of the assumption is confirmed by the verified correctness of the special relativity theory, as well as by some fairly direct evidence[2] based on observations of light from double stars.

It is clear that there must be some alteration in the usual notion of physical quantities if two observers, moving relatively to each other, are to find the same value for the velocity of light coming from a single source. The proposal that Einstein made is that the space and time variables must differ between the systems of the two observers, in a manner such that each observer will find light to have the same speed. That is, there will be a difference in clocks and measuring rods of one system, as compared with those of another system, the difference being that which is required to insure that measurements of light speed will give the same value in both systems. These differences only appear in considerations of measurements of two systems moving with respect to each other, so there is no possibility of measurements within a given system reflecting a state of absolute motion or rest.

The required changes in space and time variables are just the aforementioned Lorentz transformation equations. These equations may be derived directly from the two assumptions by considering the following physical situation. Two observers, S and S', are at a point O at a time $t=0$. S' is moving away from S, along the positive x-axis, with a velocity v. (See Fig. 5.) At the instant that S and S' are at O a light flash is emitted from a small spherical source at O. For S, the light spreads in a spherical front, moving away from him with a speed c in every

[2]References are given by Whittaker (see the preceding footnote), p. 38; discussion in P. G. Bergmann, *Introduction to the Theory of Relativity*, Prentice-Hall, New York, 1947, pp. 19-20. For a confirming observation using light from opposite edges of the sun see Bonch-Bruevich, *Dokl. Akad. Nauk. U.S.S.R.*, *109*, 481 (1956).

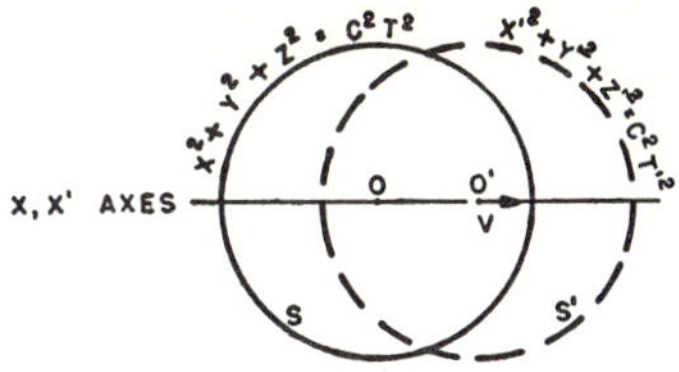

Fig. 5
A uniform spherical light signal begins to spread from O and O' when they are at the same point, at time $t=t'=0$. The system S' has a velocity v with respect to the system S, but observers at the origin O' of S' and the origin O of S must both see a spherically spreading light signal.

direction. After a time t, the spherical wave front of the light is at a distance ct from S, in every direction, and the front may be described in terms of variables of the Cartesian coordinate system of S by the equation

$$x^2+y^2+z^2=c^2t^2. \tag{1}$$

For the observer S' the wave front must also be spreading uniformly in every direction, with a velocity c, since the assumption of the relativity theory is that any light wave has a velocity c with respect to an observer, regardless of what the velocity between observer and light source may be. If we take it that the spreading spherical wave front is a unique physical entity, there must be some difference in the space and time variables of S' as compared with those of S, if S' is also to find the wave front spreading out in a uniform sphere from his coordinate origin, O'. At a time t', S' should find the wave front to be at a distance ct' from him, in any direction. In terms of the variables of the S' coordinate system, the wave front at t' may be described by the equation

$$x'^2+y'^2+z'^2=c^2t'^2. \tag{2}$$

There are relations which must exist between the variables of Eqs. (1) and (2), if both equations are to be valid for the situation as described, that is, for S' moving with a velocity $v=v_x$ with respect to S. These relations, which constitute the Lorentz transformation equations, are as follows:

$$x'=\frac{x-vt}{\sqrt{1-v^2/c^2}} \tag{3}$$

$$t' = \frac{t - vx/c^2}{\sqrt{1 - v^2/c^2}} \tag{4}$$

$$y' = y \tag{5}$$

$$z' = z. \tag{6}$$

The derivation of the Lorentz transformation equations may be found in any textbook discussion of special relativity theory, and we shall not present it here. The relations between (x,y,z,t) and (x',y',z',t') may also be found for the case in which the velocity between S and S' is in a direction not described as along the x-axis, but we shall have no need for this general form of the Lorentz transformations. The reader may find by algebra that the transformations (3) and (4) can also be expressed as follows:

$$x = \frac{x' + vt'}{\sqrt{1 - v^2/c^2}} \tag{7}$$

$$t = \frac{t' + vx'/c^2}{\sqrt{1 - v^2/c^2}}. \tag{8}$$

We note that there is no transformation of space variables in the y and z directions, if the relative motion is only in the x direction; and also, that for velocity v negligible compared with c the transformations reduce to

$$t = t' \tag{9}$$

$$x' = x - vt, \text{ or } x = x' + vt'. \tag{10}$$

The identity time transformation, (9), is of course the usual pre-relativistic one, in which it is assumed that there is a common, identical time for all coordinate systems. Eqs. (10) are the usual, non-relativistic equations for the location of point x, of S, in system S' at time t', or for the location of point x' in the system S at time t. If x is the origin O, and $x' = 0'$, the equations are simply the relation: distance=velocity $\times$ time. Finally, we point out that the transformations have the important property that if any one of Eqs. (3), (4), (7), or (8) describes a transformation between systems S and S', and if there is a similar transformation between relatively moving systems S' and S'', then there is also a similar transformation between S and S''. This property results because the Lorentz transformations are what the mathematician technically calls a *group* of transformations.

2. TIME AND FREQUENCY VARIABLES.

We shall be particularly concerned with the Eqs. (4) and (8) of the preceding section. Before discussing their use in describing physical events we shall, however, consider the physical basis of the time variable in relativity theory.

The discussion of the time concept given in the opening chapter of this book is in the spirit of the relativity-theory point of view, for in that discussion time was presented as a concept based on properties of physical systems and physical events. In the pre-relativity view, in contrast, time was considered to be a forward flow of events which is everywhere uniform, and which is not derived from physical events. Rather, there was the implicit notion that time is moving forward by its own nature, so that physical events may be referred unambiguously to the time variable. The relativistic view, and the view put forth in this book, is that time arises from physical events.

Specifically, we designate some suitable mechanism, as has been discussed in Chapter I, to be a clock. Time is then determined by the physical processes of that clock. We have also pointed out that we choose for a clock a process which is uniform in its cycles or other fiducial behavior, with respect to other processes. We can so choose a clock process because Nature does present us with a wide array of processes which bear consistent frequency relations to each other.

Suppose that a cyclical process has been chosen to be a clock, and we determine that n cycles of the process constitute a unit of time, say a second, where n is a definite, prescribed number. Further, let us assume that we have several of these clocks, all of similar physical construction and all characterized by having the same tempo, as we observe them in the laboratory. (By "having the same tempo" we mean that all of the clocks simultaneously go through the same number of cycles.)

We now carry out an experiment whereby one of the clocks moves with a high velocity with respect to the laboratory. It may then be found that the moving clock has n' cycles of its process while a similar rest clock in the laboratory has only n cycles, where $n' \neq n$. The time of the moving clock is different then from that of the rest clocks. We do *not* say that the moving clock now has n' cycles in its second of time while previously it had n cycles in its second. Instead, we do say that to the laboratory observer time has changed for the moving clock, because the rate of its physical processes has changed, and it is these physical processes which determine the time of the clock. If we said that the clock now has n' cycles in an interval in which before there were n cycles, we would be assuming that there is a unit of time, a second, which is defined over and above the physical processes of the clock, whereas in fact we have no way of defining time intervals except by reference to physical processes. To an observer at rest *with respect to* the moving clock, the tempo of the clock processes is just the same as the tempo of a similar clock at rest in the laboratory is to a laboratory observer. For both observers, if they have a common unit of time, n cycles of the clock at rest with respect to them will constitute one second. If it were otherwise—if the n' cycles of the moving clock constituted a second for the observer at rest with respect to it—we would be assuming an effect of uniform motion that could be detected entirely within the moving system; special relativity theory, and experience, deny the possibility of any such effect.

We may legitimately say that now the moving clock has n' cycles while the laboratory clock has n cycles, if we are able properly to compare the moving clock and the similar laboratory clocks at the first cycles and again at the nth, n'th cycles. And, if n' is greater than n, we may properly say that *for a laboratory observer* the moving clock is running faster, or at a faster tempo, than are the laboratory clocks. But always we must base our statements about time interval or tempo on physical processes.

If one of the laboratory clocks goes through N cycles, we say that a time interval of $N/n=t$ seconds has passed. Suppose that a clock, physically similar to the laboratory rest clocks, is in relative uniform motion with respect to the laboratory coordinate system; if that moving clock is observed to go through N' cycles, a time interval of $N'/n=t'$ seconds has passed. If N and N' can be shown to have simultaneous first cycles and Nth, N'th cycles, the time interval t for a stationary clock process has been transformed to a time interval t' for a moving clock process. The time transformation results for just the reason that there has been an effect of relative motion between two systems on the physical processes in one system as observed in the other system.

The *frequency* of a process is the number of cycles or other complete occurrences of the process in a standard unit of time. Suppose we have a fixed reference clock in the laboratory, and as before, n cycles of this clock process constitute a standard second. Suppose m cycles of some other process are observed to occur in one second—that is, during the interval of n cycles of the standard clock. We then say that the process has frequency $\nu=m$ per second. Suppose next that a process, physically similar to the one of frequency m per second, is observed from a coordinate system with respect to which the process is moving. The clock of n cycles per second is at rest in the coordinate system—which might well again be a physicist's laboratory—and with respect to this clock the moving process is observed to have m' cycles in a second. We then say that the frequency of the process is $\nu'=m'$ per sec. In general, $m'\neq m$. Relative motion has resulted in a change of frequency of the process.

It is important to note that both time interval, t, and frequency, ν, are proportional to the number of cycles of a process. If as a result of relative motion N' or m' increases with respect to the cycles of a reference clock at rest, t' or ν' becomes larger. If we wish to assign a frequency to a given process and also to use the process as a measure of time interval, we may say that ν is linearly proportional to t, or vice versa. These relations follow because both frequency and time interval are proportional to the number of cycles of the process.

We may symbolically show the validity of the direct proportionality between ν and t as follows. First, we take the process of interest and also a reference clock of n cycles per sec. to be at rest with respect to the observer. Suppose there are M cycles of the process while there are N cycles of the clock. Then frequency $\nu=m=M/t=M(n/N)$ per sec., since $t=N/n$. If the process is used as a clock, there is some number $\bar{n}$ for the process, such that there are $\bar{n}$ cycles in a second, just as there were n clock cycles in a second. Then $t=M/\bar{n}$, and both ν and t are proportional to M, so that any change in M will equally affect ν and t. Also, we see that $\nu=(\bar{n}n/N)t$, since $M=\bar{n}t$. The constant $\bar{n}$ must be regarded as independently defined, just as is n, if the process is to be used as a clock; and changes in M are changes in t, since the process defines ("measures") a time interval.

Suppose now that the process is in motion with respect to the observer and his rest clock of n cycles/sec. He may now find that there are M' cycles of the process while there are N cycles of his clock. Therefore, $\nu'=m'=M'/t=M'(n/N)$. If the moving process is used as a clock, the time interval which it indicates is $t'=M'/\bar{n}$. So again, both ν' and t' are proportional to M'. Further, again we see that $\nu'=(n\bar{n}/N)t'$, since $M'=\bar{n}t'$. The proportionality constant, $n\bar{n}/N$, is the same between ν' and t' as between ν and t. The quantities n and $\bar{n}$ are constants because they are used to define the units of time respectively for the reference clock and for the process in question, and N is constant because it is the number of clock cycles with respect to which the numbers M and M' are determined.

It is commonly taken that if the frequency of a process changes, say from ν per sec. to ν' per sec., that we do not have thereby any change in time measure but only in the number of cycles of the process occurring in a second. This point of view is proper if we are assuming a constant, fixed measure of time for all events. In ordinary non-relativistic physical considerations such an assumption is generally justified. But in relativistic considerations we say that relative motion affects the time of processes; this means that the processes *as measures of time* are affected, and therefore that t as well as ν is affected, in the manner stated in the preceding discussion. To assume otherwise would be to assume that there is a common time variable for all physical processes; the theory and experimental confirmation of special relativity contradict this assumption, because they show us that there are processes whose time variable, as related to the processes themselves, is transformed as a result of relative motion with respect to an observer.

3. THE DOPPLER EFFECT FOR LIGHT.

The effect of relative motion on sound waves coming to a listener from a moving source is well known. For example, the frequency and hence the pitch of a locomotive whistle is higher when the train is approaching than when it is moving away. There is a similar frequency effect for light waves, and the special theory of relativity leads to an equation for the effect. Experiment has confirmed that the effect is as described by the relativistic equation.

In order to derive the equation we shall consider light to be observed in a system S, and to have a frequency ν and a wave length $\lambda = c/\nu$ in that system. The source of the light may or may not be at rest in S. We take the path of the light to be along a line r that makes an angle α with the positive x-axis of the coordinate system S. (See Fig. 6.) At $t=0$ a light wave-front is at the origin of S; then $r=ct$ gives the position of that wave-front at time t. The x-coordinate of r is given by $x=ct \cos \alpha$.

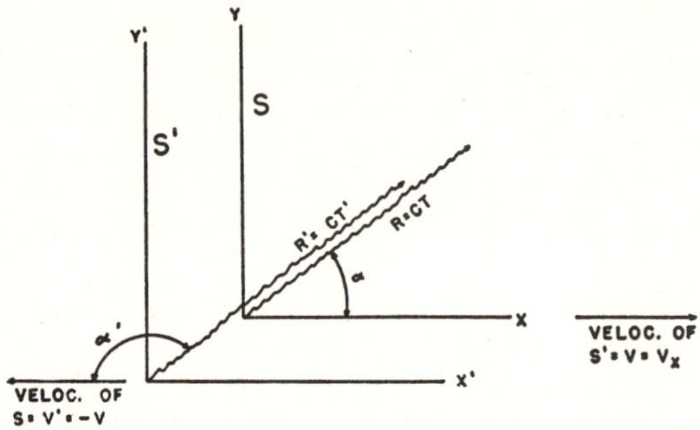

Fig. 6
S and S' are relatively moving coordinate systems from which observations are made on a light beam from a given source. A light ray from the source is represented by the vector $r=ct$ for the S observer, and by $r'=ct'$ for the S' observer.

A second coordinate system, S', moves with respect to S with a velocity v, along the x-axis of S'. To an S observer, the velocity of S'

is in the positive direction of the x-axis. We properly define α as the angle between the direction of the light ray and the velocity vector of the moving system. For an S observer, the x-axis and v have the same direction.

In S' the light has a frequency $\nu'=c/\lambda'$. We take the light front that passes through the origin of S at $t=0$ to pass through the origin of S' at $t'=0$. Then $r'=ct'$ describes the location of that front for the S' observer. For him, system S is moving along his negative x-axis with a velocity v'; $v'=-v$, where v is the velocity of S' as observed in S. The angle α' is between the vector of the light ray and the velocity vector of the moving system, both as observed in S'. For S' observers, the x-component of r' is given by $x'=r' \cos(\pi-\alpha')=-r' \cos\alpha'=-ct' \cos\alpha'$.

We now wish to compare the light-wave frequencies ν and ν' as observed from S and S'. According to the Lorentz transformations we can expect ν and ν' to differ because the times of processes will differ as they occur in S or S'. Eq. (4) of section 1 of this chapter gives us the time transformation:

$$t'=\frac{t-vx/c^2}{\sqrt{1-\beta^2}},$$

where (as we shall now usually do in writing the Lorentz equation) the ratio v/c has been written as β.

In making application to the light ray we may substitute $ct \cdot \cos\alpha$ for x, so that the equation becomes

$$t'=\frac{t[1-\beta\cos\alpha]}{\sqrt{1-\beta^2}}.$$

However, we have seen (section 2) that ν and ν' are proportional to t and t', with the same proportionality constant. Hence,

$$\nu'=\frac{\nu[1-\beta\cos\alpha]}{\sqrt{1-\beta^2}}. \tag{1}$$

This equation is a Doppler equation for the effect of velocity on frequency. The use of Eq. (8) of section 1 and of $x'=-ct'\cos\alpha'$ gives the frequency transformation:

$$\nu=\frac{\nu'[1-\beta\cos\alpha']}{\sqrt{1-\beta^2}}. \tag{2}$$

Equations (1) and (2) are not however in a very useful form. Eq. (1) gives the frequency ν' for an S' observer, but in terms of the angle α, which would be measured by an S observer; likewise (2) gives ν in terms of α'. We may readily rearrange the equations into the following:

$$\nu=\frac{\nu'\sqrt{1-\beta^2}}{1-\beta\cos\alpha} \tag{3}$$

$$\nu'=\frac{\nu\sqrt{1-\beta^2}}{1-\beta\cos\alpha'}. \tag{4}$$

Eq. (3) now gives the frequency which an S observer will find for light which makes an angle α with the velocity vector for S'; this light has a frequency ν' in S'. Similarly (4) gives ν', for an S' observer, in terms of α' and the frequency ν in S.

If the light source is at rest at the coordinate origin of S', we may write $\nu'=\nu_0$. Also, if the source is moving directly toward or away from the S observer, so that α is π or 0, we have $\cos\alpha=\pm 1$. Under these conditions Eq. (3) becomes:

$$\nu=\frac{\nu_0\sqrt{1-\beta^2}}{1\pm\beta}. \tag{5}$$

Eq. (5) tells us that if the light source is moving directly toward the S observer ($\alpha=0$), $\nu=\nu_0\sqrt{1-\beta^2}/(1-\beta)$, and neglecting the generally small $\sqrt{1-\beta^2}$ factor, the effect of motion is a frequency increase. If the light source is receding directly from the S observer ($\alpha=\pi$), $\nu=\nu_0\sqrt{1-\beta^2}/(1+\beta)$ and the first order effect is a decrease in frequency.

A derivation of the Doppler equation for light waves, without any consideration of the Lorentz transformation on x and t variables, leads to equations of the form,

$$\nu=\frac{\nu'}{1-\beta\cos\alpha}. \tag{6}$$

The factor $\sqrt{1-\beta^2}$ in the relativistic Eqs. (3) and (4) is, then, the contribution of relativity theory to the Doppler equations for light. Since v^2/c^2 is small except for velocities which do approach the velocity of light, the $\beta\cos\alpha$ term of the denominator has been the important term in applications of the equations. But, we shall now see that the square-root factor has provided a possibility for direct experimental test of the physical reality of the time transformations.

An experimental test of the *relativistic* Doppler equation was first carried out by H. E. Ives,[3] aided by G. R. Stilwell, in 1938. G. Otting[4]

[3]Herbert E. Ives and G. R. Stilwell, *Journ. of the Opt. Soc. of America, 28,* 215 (1938); *31,* 369 (1941).

[4]Gerhard Otting, *Physikal. Zeits., 40,* 681 (1939). See also H. I. Mandelberg and L. Witten, *Bull. Amer. Phys. Soc., 5,* 360 (1960). The Mossbauer-effect studies of temperature dependence of γ-rays emitted from nuclei in solids also give confirmation of the relativistic Doppler effect. See B. D. Josephson, *Phys. Rev. Letters, 4,* 341 (1960), and C. W. Sherwin, *Phys. Rev., 120,* 17 (1960).

made a test in a somewhat different manner in 1939. Both experiments gave good agreement between observation and theory.

Ives and Stilwell employed as their light sources hydrogen atoms which had been accelerated to a high velocity. In their apparatus hydrogen molecules were first ionized by a spark discharge, and then accelerated in an electrical field. The ions were carried to values of v that were at the most of the order of one-hundredth the velocity of light. The observations were made on the H_β line of the hydrogen spectrum, of wave length 4861.33 Å. The wave lengths of the H_β line were measured both for H atoms moving directly away from the observer, and for atoms coming directly toward the observer. Ives and Stilwell then determined the center frequency between these two displaced lines, or, as Ives called it, the "center of gravity" of the two lines. The validity of the relativistic prediction depends on whether this center falls at the center line as determined by the relativistic Eq. (4) or as determined by the non-relativistic Eq. (6), in each case with $\alpha=0$ or π.

Specifically, the frequency of the light from atoms moving toward the observer is $\nu=\nu_o\sqrt{1-\beta^2}/(1-\beta)$, using (3) or (5), where ν_o is the frequency from H atoms that are moving only with small velocities in the laboratory. For atoms receding with velocity v from the observer, $\nu=\nu_o\sqrt{1-\beta^2}/(1+\beta)$. The mean of these two ν values is given by

$$\bar{\nu}=\frac{1}{2}\left\{\frac{\nu_o\sqrt{1-\beta^2}}{1+\beta}+\frac{\nu_o\sqrt{1-\beta^2}}{1-\beta}\right\}=\frac{\nu_o}{\sqrt{1-\beta^2}}. \tag{7}$$

Since powers of v^2/c^2 above the first will be negligibly small, Eq. (7) may be converted by a series expansion of $1/\sqrt{1-\beta^2}$ into

$$\bar{\nu}=\nu_o(1+\frac{1}{2}\beta^2). \tag{8}$$

If the non-relativistic equation (6) is used, $\nu=\nu_o/(1-\beta)$ or $\nu=\nu_o/(1+\beta)$, for light from atoms that are approaching or receding from the observer respectively. Then the mean frequency is

$$\bar{\nu}=\frac{1}{2}\left\{\frac{\nu_o}{1-\beta}+\frac{\nu_o}{1+\beta}\right\}=\frac{\nu_o}{1-\beta^2}\quad . \tag{9}$$

With neglect of powers of v^2/c^2 higher than the first Eq. (9) may be written as

$$\bar{\nu}=\nu_o(1+\beta^2). \tag{10}$$

The work of Stilwell and Ives, and also of Otting, definitely showed a frequency shift in accordance with Eq. (8) rather than (10); that is, the mean displacement of the H_β lines from the moving atoms was

$\frac{1}{2}(v^2/c^2)\nu_o$. The experimental accuracy was ample for distinguishing from the non-relativistic displacement of $(v^2/c^2)\nu_o$.

4. MOTION AND MESON DECAY TIMES.

The relativistic Doppler effect just discussed is a demonstration of the validity of the Lorentz time transformation, for a purely cyclic process, namely, light waves. Impressive further evidence for the validity of the time transformation is found in the effect of relative motion on the decay times of mu-mesons. In this case the process affected is a non-cyclic disintegration, with a directly measured time span.

Mu-mesons are one of the products of interactions of primary cosmic ray particles with atoms high in the earth's atmosphere. The mu-meson has a mass about 200 times that of an electron, has a unit (+) or a unit (−) charge, and is unstable, decaying into two neutrinos and a positron or electron. The mean life-time τ_o, for mesons not moving at high speeds, has been observed as 2.2 millionths of a second. Many mu-mesons are, however, given high kinetic energy as they are produced, and come down through the atmosphere at speeds which approach the speed of light. For these mesons we might expect an increased life-time, as a result of the relativistic time transformation. This life-time increase has been observed.

The effect of motion on the meson life-time is readily calculated from Eq. (8) of section 1 of this chapter. We re-write this equation for increments of t, t', and x' in the form:

$$\delta t = \frac{\delta t' + (v/c^2)\,\delta x'}{\sqrt{1-\beta^2}}, \tag{1}$$

where we again write v/c as β.

This new form is justified because in the derivation of the Lorentz transformations any space or time interval that we choose may be transformed. We now assume that a mu-meson is at rest at a point x'_o in S'. The system S' is then moving with respect to the laboratory observer with a velocity equal to that of the meson with respect to the laboratory. Since the meson is at a fixed point in S', $\delta x'=0$ and (1) becomes,

$$\delta t = \frac{\delta t'}{\sqrt{1-\beta^2}}. \tag{2}$$

We may take $\delta t'$ as the mean life-time, τ_o, of a mu-meson to a rest observer, and δt as the mean life-time τ to an observer with respect to whom there is relative velocity v. Hence, (2) gives us,

$$\tau = \frac{\tau_0}{\sqrt{1-\beta^2}}. \quad (3)$$

Equation (3) has been confirmed for mu-mesons by several groups of workers.[5] Mesons coming through a given solid angle may be detected by means of a "meson telescope." This detector is a group of absorbers and Geiger-Muller counting tubes, so arranged as to signal the passage of a meson along the axis of the "telescope." Measurements of the mu-meson intensities may be made at two different altitudes, say z_1 and z_2, with the same counting apparatus. If the measured intensities were the same at z_1 and z_2, no decay would be indicated, since all the mesons at the higher altitude, z_1, would then be getting down to z_2. But what is found is that there is a smaller intensity at z_2, and therefore there is a disappearance of mesons between z_1 and z_2.

This disappearance is not however merely a result of radioactive decay, since there is also an absorption of mu-mesons by matter. Some of the mesons travelling downwards between z_1 and z_2 are absorbed by the atmosphere. It is possible however to compensate for the air absorption between z_2 and z_1 by placing above the counting apparatus at z_1 a solid absorbing block, of a thickness equivalent in absorbing power to the power of the atmosphere between z_2 and z_1. Then the mesons reaching z_2 and z_1 have passed through the same equivalent absorbing layers of matter, and the decrease in recorded intensity at z_2 is a measure of the radioactive decay between z_1 and z_2. From the intensity measurements, the mean life-time τ may be calculated. It is necessary that there be counting only of mu-mesons in a given energy range; by use of suitable absorbers such a selection of mesons may be achieved.

Specifically, we will consider the experiments carried out by Rossi and his co-workers, at Echo Lake and Denver in Colorado. The elevation of Echo Lake, corresponding to z_1, is 1624 meters greater than that of Denver, corresponding to z_2. Measurements at Echo Lake were made with an additional iron absorber of 200 gm./cm.^{2}, which is approximately equivalent in absorbing power to the 1624 meters of air between Echo Lake and Denver. One set of measurements was of mu-mesons which had energies of the order of 400 million electron volts at Denver; the velocities of these mesons would be very close to the velocity of light.

We may express the mean distance traversed by the mesons as $L = v(\tau_o/\sqrt{1-\beta^2})$, or, $L = p\tau_o/m_o$ (4)

[5]W. M. Nielsen, C. M. Ryerson, L. W. Nordheim, and K. Z. Morgan, *Physical Rev. 57,* 158 (1940); G. Bernardini, B. N. Cacciapuoti, E. Pancini, O. Piccioni, G. C. Wick, *idem., 60,* 910 (1941); G. Bernardini, *Zeits. f. Physik, 120,* 413 (1942); B. Rossi and D. B. Hall, *Physical Rev., 59,* 223 (1941); B. Rossi, K. Greisen, J. C. Sterns, D. K. Froman, and P. G. Koontz, *idem., 61,* 675 (1942); D. W. Seymour and W. F. G. Swann, *Journ. Franklin Inst., 258,* 277 (1954). Also, Bruno Rossi, *High Energy Particles,* Prentice Hall, New York, 1952, sec. 4.7.

where v is the meson velocity, p is the momentum, defined by $p = m_o v / \sqrt{1-\beta^2}$, and m_o is the rest-mass of the mu-mesons. Rossi used an effective momentum p, intermediate between the initial and final momenta. Suppose that n_1 is the number of mu-mesons entering the apparatus at Echo Lake per minute, and n_2 the number at Denver. The probability of survival for mesons, in travelling from z_1 to z_2, is n_2/n_1. We may write this probability as $w_{12}=n_2/n_1$.

In order to relate the τ_o of Eq. (3) to observed quantities we must have a relation between L and w_{12}. This we readily find as follows. If there are n mesons at point z, the decrease in number of mesons, $-dn$, as the mesons traverse a distance $-dz$, is given by $-ndz/L$. This follows because L is the distance in which all n of the mesons will on the average cease to be mesons. In equation form, $dn=ndz/L$, or,

$$\int_{n_1}^{n_2} \frac{dn}{n} = \int_{z_1}^{z_2} \frac{dz}{L}.$$

Integration gives us, $\log_e(n_2/n_1)=(z_2-z_1)/L$, or,

$$\log_e w_{12} = -\frac{z_1-z_2}{L}. \qquad (5)$$

Rossi's measurements, for the mu-mesons in the energy range indicated, gave for n_2/n_1 a value of 0.70, with an experimental uncertainty of about 0.03. Then, since z_1-z_2 is 1624 meters, we find from (5) that $L=1624/(-\log_e 0.70)=4.5\times10^3$ meters. We may now solve Eq. (4) for τ_o, writing the equation as $\tau_o=m_o L/p$. The 400 million electron volt mu-mesons have an effective momentum p of about 2.7×10^{-14} gm.cm./sec., and the rest mass of mu-mesons, taken as 200 electron masses, is 1.8×10^{-25} gms. We find then,

$$\tau_o = \frac{1.8\times10^{-25}\times4.5\times10^5}{2.7\times10^{-14}} = 3.0\times10^{-6} \text{ secs.}$$

This result is somewhat higher than the currently (1957) accepted value of 2.2×10^{-6} sec. for τ_o, obtained by direct measurement of the decay of mu-mesons at rest. Later workers, using methods similar to the one we have described, have found a τ_o value closer to 2.2×10^{-6} sec., but in view of the uncertainties involved in the value of p, even the value obtained by Rossi is satisfactory confirmation of the time transformation.

It should be noted that if L were taken as $v\tau_o$, instead of as $v(\tau_o/\sqrt{1-\beta^2})$, the average range of the mu-mesons would be only $c\tau_o=3\times10^{10}\times2.2\times10^{-6}=660$ meters, since v is approximately equal to c for these energetic mu-mesons. Without the "time dilatation" virtually none of the mesons observed at Echo Lake would have been counted in Denver, 1624 meters lower in altitude.

The observations of the change of mu-meson lifetime with relative motion are indeed striking, and knowledge of the effect must have a major role in our gaining an understanding of natural time processes. A first question that might arise is: Suppose a group of mu-mesons were formed in empty space; with respect to what body would they have the relative velocity that would determine their mean life-time? The answer must be: with respect to the atoms with which the decay electrons of the mesons interact. There hardly can be any alternative answer, because we have no evidence for any kind of physical reference system of empty space by which we can determine absolute velocity in space—this is the lesson which physicists learned in discarding the "ether" of pre-relativity theory days.

The mu-mesons for which we observe a Lorentz transformation of life-time are coming through the atmosphere, and their decay electrons interact with the atmospheric atoms. These atoms have speeds, with respect to the earth, which are negligibly small in contrast with the speed of light, and therefore may be taken as being at rest with respect to the earth. Hence, the mu-mesons show the time change that is to be expected as a result of their velocity with respect to the earth. Mesons that have been "caught" in absorbing blocks in the laboratory give up their decay electrons to counting devices with respect to which they are at rest, from a relativity-theory point of view, and therefore these mesons show a rest mean life-time, τ_0.

It is helpful to think of the relativistic meson life-time effect as another kind of Doppler effect. Just as light from a given source will have altogether different frequencies, wave-lengths, and energies to observers with different velocities, so will mesons of a given group have different life-times for different observers, depending on the relative velocity of each observer's detecting apparatus. As with the optical Doppler effect, there is no occasion for paradox. Each photon from a light source has an observed frequency dependent upon the velocity of the observer with respect to the source, and a given photon can interact only with one observer. Likewise, the life-time of a meson is determined—on the average—by the relative velocity of the material with which its decay electron interacts, and each meson decays with respect to some given material aggregate. There is no basis for two different and contradictory frequencies for a given photon, because only one observer can interact with the photon and determine its frequency. Likewise with mesons; only one observer can detect the electrons emitted from a group of mesons and thereby determine the mean life-time of the group. The electrons counted by this observer might pass through the counters, on to a second observer moving rapidly with respect to the first. But the mesons would have their primary decay interaction with respect to the counters of the first observer, and we should expect the meson life-times to be as calculated from their velocity with respect to that observer.

We do not today know in what precise manner relative motion determines the frequency of a photon or the half-life of a meson. But it is clear that an interaction between photon or meson and a single observer is necessary for frequency or life-time determination. Therefore we cannot properly refer to a photon or meson as a public clock, which may be read by any number of observers with differing velocities. In fact, the photon or meson ceases to exist as its frequency or half-life is read by a single observer. This non-public nature of mesons, as clocks, will be important in our discussions of relativistic time paradoxes in Chapters VII and VIII.

A final point may be made about the mu-meson as a clock. In the determination of the *rest* life-time τ_0, mesons of varying energies are brought to approximate rest in an absorbing block.[6] The numbers of decays n after various times of arrival Δt_n are observed, and a plot of n against Δt_n gives a line whose slope is the negative of the decay constant. This is the same procedure as that by which the decay constant of any radio-active material may be found, and the mean life-time bears a constant ratio to the decay constant.

In some of the determinations of τ_0, the mu-mesons have been moving with speeds close to that of light before reaching the absorbing block, but the past motion is found to have no effect on the observed τ_0. This fact is in accordance with our observation that the life-time is determined by the motion of the meson, at the instant of decay, with respect to the matter with which its decay electron interacts. Similarly, *past* relative motion of an observer has no effect on the frequency which he finds for a photon of light by interacting with it.

5. DE BROGLIE WAVES

The relativistic Doppler equation results from application of the Lorentz transformations to light, which has the same velocity in every coordinate system. The equation relating meson decay time and relative velocity required an application of the transformations to a physical process located effectively at a fixed point in a coordinate system. Another application of the transformations can be made to a process not at a point but extended over a distance, and yet at rest in a given coordinate system. This application, taken with the energy-related frequency of matter, leads to the equations for de Broglie waves.

---

[6]B. Rossi and N. Nerenson, *Physical Rev.*, *62*, 417 (1942); *64*, 199 (1943); M. Conversi and O. Piccioni, *idem.*, *70*, 859 (1946); H. K. Ticho, *idem.*, *74*, 1337 (1948); L. W. Alvarez et al., *idem.*, *77*, 752 (1950); R. E. Bell and E. P. Hintes, *idem.*, *84*, 1743 (1951).

We write Eq. (4) of section 1 of this chapter, for time and space increments:

$$\delta t' = \frac{\delta t - (v/c^2)\,\delta x}{\sqrt{1-\beta^2}}. \tag{1}$$

A space interval $\delta x'$ is in the coordinate system S', and $\delta t'$ is the time interval associated with this interval, in S'. In S, with respect to which S' has a velocity $v = v_x$, the length $\delta x'$ is observed to be δx, and the interval $\delta t'$ is transformed to δt. Now, in general an observer does not find that a time interval is associated with a stationary length; that is, events or processes which are synchronized so as to present the same time characteristics to a stationary observer, when they are at a common point, are found to continue to present the same time characteristics when they are situated some distance apart (but still at rest to the observer). Two synchronized physical processes, for example, would not be expected to show to rest observers any kind of phase difference—any temporal lag—as a result of being spatially separated. We may then, because a motionless space extension does not involve a time interval, assert that the interval $\delta t'$ associated with $\delta x'$ is zero. This is only to say that an observer in S' finds zero time interval associated with the S' length $\delta x'$.

We now see however, from (1), that the intervals δx and δt, which are the space and time intervals associated with $\delta x'$ as observed from S, will be related by the equation:

$$\delta t = (v/c^2)\,\delta x. \tag{2}$$

This result follows at once from (1) on setting $\delta t' = 0$. The important result stated by (2) is that two events, simultaneous in S' and separated by a length $\delta x'$, observed in S to be a moving length δx, have a time interval $v\delta x/c^2$ between them in S. The existence of a spatial separation between simultaneous events results in a temporal separation if the events are observed from a system that has a velocity along the line of spatial separation of the events; or, speaking rather loosely, we might say that a moving length has a time interval associated with it.

We expect that events of a physical process, spatially separated, which are simultaneous in one system will not be so as observed from a relatively moving system. Specifically, we want to apply this expectation to aspects of a moving elementary particle, observed over a spatial interval. In accordance with our physical observation that spatial separation in a rest system does not imply a time difference in a physical process, we assign simultaneity to a pair of events in the moving particle system S', separated by distance $\delta x'$. But how may laboratory observers interact with spatially separated events of a moving system? One type of situation in experimental physics in which such an interaction occurs is in diffraction and interference experiments. In these situations, different parts of a photon or particle are brought together after the parts have travelled over different path lengths from a source. The "path-length difference" of an interference observation in laboratory system S is in effect a length interval, δx, because it represents a separation of parts of the extended waves which are observed. The waves are of a moving process, and δx is the separation in terms of the coordinates of the system S, here representing the fixed laboratory coordinate system. As an example, we sketch in Fig. 7 a wave process,

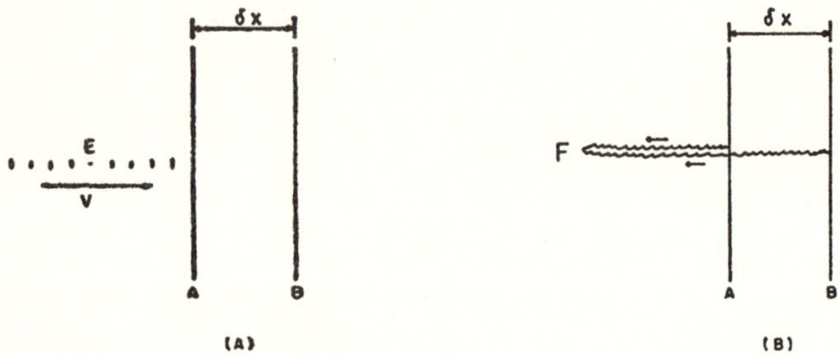

Fig. 7
7 (A): An electron E approaches crystal layers A and B in the laboratory system with velocity v.
7 (B): Waves reflected at layers A and B are brought together at F. These waves are elements of the moving electron wave process separated by a distance δx.

associated with an entity, as say an electron, which is at rest in system S'. The entity and system have a velocity v with respect to two layers, a and b, separated by a distance δx, of a crystal in the laboratory system S. The wave process is reflected partially at layer A and partially at B, and then brought to focus at F for an observation of interference. Two wave elements which come together at F are elements of the original wave process of S' that are separated by a distance δx. ($\delta x'$ in S'.) Further, the process is moving at a velocity v with respect to S, so we are justified in saying that observation is being made of a *moving* entity extended over a distance δx.

With this experimental application in mind, we apply Eq. (2) to a wave process. We consider waves which have frequency ν and wavelength λ to an S observer; since Eq. (2) is to apply we require that in the rest system S' the process presents a zero time interval in association with the space interval. In an S interval there are $\delta x/\lambda$ waves. In association with this space interval there is a time interval $v\delta x/c^2$, by (2); this time interval, by the relativity transformations, is one which should be physically manifested in any temporal characteristics of the wave process under observation. Since the waves have a frequency ν, in the time interval there should be $\nu\delta t = \nu v \delta x/c^2$ waves. The interval must contain the same number of waves, whether the number is determined from the wave-length or the frequency of the waves, so $\delta x/\lambda = \nu v \delta x/c^2$. Division by δx and re-arrangement give

$$\nu\lambda = c^2/v. \tag{3}$$

The product $\nu\lambda$ is the velocity of the individual waves, and it should not be confused with the velocity v of the entity or process which is giving rise to the waves.

Alternatively, Eq. (3) may be derived by writing as an equation for the varying amplitude $u(x,t)$ of the wave: $u(x,t) = A \sin(\nu t - x/\lambda)$, where A is a constant. Then, in accordance with the usual property of this description of a wave, the wave velocity is $\nu\lambda$, and $\nu t = x/\lambda + \phi$, where ϕ is a phase constant. This last equality requires the relation $\nu\delta t = \delta x/\lambda$ between the intervals δt and δx. Eq. (3) then follows on substitution of $\delta t = v\delta x/c^2$.

The de Broglie equation is a consequence of (3) and of an ascription[7] of a frequency ν, given by $E = h\nu$, to a mass of energy $E = mc^2$, where h is Planck's constant. Substitution of $\nu = E/h$ and $E = mc^2$ into (3) gives us,

$$\lambda = h/mv, \tag{4}$$

which is the de Broglie wave equation in one dimension. The equation tells us that a mass m, at rest in S', will have a wave-length λ associated with it for an S observer. Further, we see that the wave-length becomes infinitely long for $v = 0$. For a given mass there is no limit on the smallness of λ; the velocity v of the moving particle must always be less than c, but the mass m increases with velocity (see section 6 of this chapter) and may go to any value for which sufficient energy can be given to the particle. The "phase wave" velocity $\nu\lambda$, given by (3), is greater than c, since $v < c$, and becomes infinite for $v = 0$. This wave velocity however is not observable, just as phase waves of ordinary light—whose phase-wave speed may exceed c—are not observable. It is a property of de Broglie waves—we shall not give the derivation here—that their *group velocity* is the same as the (observable) velocity v of the mass m that is giving rise to the waves of wave-length λ.

The de Broglie equation has been abundantly confirmed by interference experiments with electrons, neutrons, and even larger mass units such as hydrogen molecules and helium atoms.[8] "Electron dif-

---

[7]L. de Broglie, *Ann. de phys.*, *3*, 22 (1925).

[8]The older experiments are discussed in G. F. Meyer, *The Diffraction of Light, X-rays, and Material Particles*, Univ. of Chicago Press, 1934. For a discussion of the wave properties of neutrons see R. S. Shankland, *Atomic and Nuclear Physics*, 2nd ed., Macmillan, New York, 1960, pp. 383-390.

fraction" is now a standard tool in academic and commercial laboratories for the study of the crystalline structure of the surface layers of solids. In all of the observations which directly demonstrate the reality of de Broglie waves, the experimental procedures are similar to those in which electromagnetic wave-lengths are measured, in that there is an interference between different parts of the "matter wave" which have been separated by traversal of different spatial distances.

The de Broglie waves are a further confirmation of the special relativity time transformation; but also, the observation of the waves confirms the physical validity of de Broglie's association of an intrinsic frequency ν with matter. Planck's original work with quantum theory required that electromagnetic radiation be in quanta of energy $h\nu$ in interaction with matter, and the description of the photo-electric effect by Einstein and of atomic energy levels by Bohr utilized and confirmed this requirement. The extension by de Broglie of the intrinsic frequency to matter—that every mass of energy mc^2 has a frequency ν equal to mc^2/h—was a superb insight that has been the basis of an immense growth of our physical knowledge.

An intuitive understanding of the relation between de Broglie waves and the relativistic time transformation may be gained in the following way. If an ordinary wave process is considered at a space point, the passage of time brings a cyclical alteration of the wave amplitude at the point. That is, the phase of the wave changes from peak to zero to trough to zero to peak and so on, as time passes. Now, for a *particle* at rest to its observer or to any diffraction apparatus there is no time interval associated with a length interval, and all spatial parts of the "wave" of the particle are at the same phase. In other words, the wave-length is infinite. But when the particle is moving with respect to a diffraction apparatus, a space element δx requires a time separation $v\delta x/c^2$ in the interacting wave process. This separation appears as a *phase difference* in the wave at different space points, with the result that a complete wave now fits into a space interval $\lambda = c^2/v\nu$. The transformation required by Eq. (2) from a zero to a non-zero time interval between wave events manifests itself as physical change in the wave characteristics of the particle.

It is interesting to note that in terms of the time variable defined by the de Broglie wave process, there is time for the parts of a wave train to "draw together" when the particle interacts as a particle rather than as a wave with an entity at rest to the observer. Each element of the wave, δx distant from a chosen $x=0$ point of the wave, is separated by a time interval $v\delta x/c^2$ from the wave element at $x=0$. We suppose that the wave is to interact as a particle at that point. Moving with its wave velocity of c^2/v, the part of the wave at δx reaches $x=0$ in the time interval of separation, $v\delta x/c^2$.

Finally, we emphasize that the de Broglie waves of a group of particles are different for observers of different relative velocity. There is of course no such quantity as *the* wave-length of an electron; an electron can assume any of an infinite number of wave-lengths, depending on what is its relative velocity with respect to the apparatus with which its wave-length is observed. The de Broglie wave is—as with the meson decay time—a "Doppler-like" effect. A photon potentially has any wave-length, depending on the velocity of the observer with respect to the source of the wave. Likewise the material particle has a wave-length that is determined by its velocity relative to the observer. And, just as with photons and mesons, there is no paradox about a particle's having different wave-lengths for different observers; the particle, like a photon or meson, can interact only with one apparatus, and therefore can exhibit only one wave-length to that apparatus.

6. VELOCITY AND MASS-ENERGY.

The effect of relative velocity on the mass-energy of a particle is one of the best known results of the special theory of relativity. The equation for the effect may be derived in a number of ways, and indeed, it was shown by G. N. Lewis[9] that the equation must obtain for interactions between electromagnetic radiation and mass, even without any regard for considerations of the special theory of relativity. A very simple derivation is based on a direct consideration of the effect of relative velocity on the intrinsic frequency ν of any mass, and we shall present this derivation. The validity of the de Broglie equation justifies our use of the fundamental mass-frequency, and the derivation will show the mass increase as an effect of relative velocity on the time of a process.

We consider a particle of rest mass m_o to be fixed at the point x_o' of the coordinate system S'. The mass has an intrinsic frequency ν_o associated with it, such that $h\nu_o$ is an energy E_o of the mass; at this point we assume that we have no other knowledge concerning what kind of an energy E_o is. The mass m_o is observed from a coordinate system S, with respect to which S' has a velocity $v=v_x$. The mass is denoted by m for an observer at rest in S, and for this observer the intrinsic frequency of the mass is ν and the associated energy $h\nu$ is E.

Now, by Eq. (8) of section 1, the same basic equation that we used in section 4 in considering a meson at rest at x_o, the time variable of any process of the mass at rest at x_o' will be related for S' and S observers by the equation

[9] G. N. Lewis, *Phil. Mag., 16,* 705 (1908); *Science, 30,* 84 (1909).

$$\delta t = \frac{\delta t_o - (v/c^2)\,\delta x'}{\sqrt{1-\beta^2}}, \tag{1}$$

where we have written the time interval for the S' observer as δt_o rather than as $\delta t'$. Since the mass is at rest in S', $\delta x'=0$. Hence (1) becomes,

$$\delta t = \frac{\delta t_o}{\sqrt{1-\beta^2}}. \tag{2}$$

Since frequencies in S and in S' are in the same ratio as are time intervals in S and S', as was shown in section 2, we may write (2) as

$$\nu = \frac{\nu_o}{\sqrt{1-\beta^2}}, \tag{3}$$

where ν and ν_o are the intrinsic frequencies associated with m and m_o. By the relation $E=h\nu$ we substitute E and E_o for ν and ν_o in (3) and find,

$$E = \frac{E_o}{\sqrt{1-\beta^2}}. \tag{4}$$

The difference in the energy of the mass at x_o' as observed by S and S' observers is $\Delta E = E - E_o$. By Eq. (4),

$$\Delta E = E_o\left(\frac{1}{\sqrt{1-\beta^2}} - 1\right).$$

In the case of small velocities we know that ΔE is the non-relativistic expression for kinetic energy, $\frac{1}{2}m_o v^2$. And, in the limit $v=0$ the Lorentz transformations become the transformations of non-relativistic physics. Hence we may write as a rigorously valid equation,

$$\lim_{v\to o}\left\{\Delta E = E_o\left(\frac{1}{\sqrt{1-\beta^2}} - 1\right) = \frac{1}{2}m_o v^2\right\}. \tag{5}$$

After making a binomial theorem expansion of $1/\sqrt{1-\beta^2}$, $\beta = v/c$, we may write (5) as:

$$\lim_{v\to o}\left\{E_o\left(\frac{1}{2}\frac{v^2}{c^2} + \frac{3}{8}\frac{v^4}{c^4} + \frac{5}{16}\frac{v^6}{c^6} + \ldots\ldots\right) = \frac{1}{2}m_o v^2\right\}.$$

Multiplication by $2/v^2$ gives,

$$\lim_{v\to o}\left\{E_o\left(\frac{1}{c^2} + \frac{3}{4}\frac{v^2}{c^4} + \frac{5}{8}\frac{v^4}{c^6} + \ldots\ldots\right) = m_o\right\}.$$

We next evaluate the equation for $v=0$, and find

$$E_o(1/c^2) = m_o, \text{ or, } E_o = m_o c^2. \tag{6}$$

We have found, then that the energy $h\nu_o$ associated with the rest mass m_o is $E_o = m_o c^2$. Substitution of $E_o = m_o c^2$ into (4) gives us the energy E associated with m:

$$E = \frac{m_o c^2}{\sqrt{1-\beta^2}}. \tag{7}$$

The term $m_o/\sqrt{1-\beta^2}$ is the "relativistic mass" of the particle at x_o'. That is, the mass m for the observer in S is given by:

$$m = \frac{m_o}{\sqrt{1-\beta^2}}. \tag{8}$$

We have found then, in Eqs. (3), (4), and (8), that ν_o, E_o, and m_o are increased by a factor $1/\sqrt{1-\beta^2}$ when observed from the S coordinate system. By (6) and (7) the mass and energy of the particle, as observed in any coordinate system, are related by the invariant factor c^2, and so we refer to these two properties of the particle as its mass-energy.

The experimental evidence in favor of relativistic mass-energy equivalence is so great as to place these relations beyond question. Even before Einstein's first paper on special relativity, W. Kaufmann[10] had observed an increase in the mass of electrons moving with very high velocities—or more precisely he had found a decrease in the e/m ratio of electrons, where e is the electron charge, a magnitude unaffected by relative velocity. Today, physical theory virtually everywhere makes use of the mass-energy-velocity relations, and activities such as the operation of high-energy accelerators or the utilization of nuclear energy are confirmations in daily practice of the relations.

Equations (3) and (4) may be regarded as having direct verification in the common behavior of bodies at velocities that are very small compared with the velocity of light. Since ΔE, Eq. (5), is within experimental error equal to $\frac{1}{2}mv^2$ for such small velocities, the fact that energy $\frac{1}{2}mv^2$ is required to give a velocity v to a mass m is a confirmation of the effect of relative velocity on energy. (The quantity E_o was, it is true, evaluated by Eq. (5) so as to give $\Delta E = \frac{1}{2}m_o v^2$ at the limit of $v \to 0$. Nonetheless, the form of E as given by (4) is such that the energy associated with any v is $\frac{1}{2}mv^2$, until v begins to approach c and the relativistic kinetic energy expression must be used.)

Alternatively, we may consider Eq. (4) as a statement of the inertia

[10] See E. T. Whittaker, *op. cit.*, p. 53.

of mass. We commonly say that because of inertia we must expend energy in order to set a body in motion. Eqs. (3), (4), and (5) tell us that the source of the inertia is the change in fundamental frequency that is associated with a velocity change. If ν remained constant for all relative velocities, E would also be an invariant and masses would presumably be indifferent to whether they were accelerated or not—the largest mass could then be given any desired velocity without expenditure of energy. The relation which does exist between energy, ν, and the Lorentz time transformation suggests that inertia is a direct consequence of the effect of relative motion on the time of physical processes. And, it is possible to show that the inertial properties of matter, as expressed in Newton's second law of motion, are a consequence, without additional physical assumption, of the Lorentz time transformation, the $E=h\nu$ relation, and the observed existence of de Broglie waves.[11]

Finally, we note that the effect of velocity on mass, like the other relativistic effects considered in this chapter, is a result of the variation of the time (frequency) variable for different observers. The mass of an electron, for example, potentially is an infinite number of different quantities. However, the mass of a particle is observed by interaction of the particle with a given measuring apparatus: specifically, in practice, with an electrical, magnetic, or gravitational field. Different masses are possible for a given particle, but only under different circumstances—that is, only with different velocities relative to the sources of the measuring field. There is no possibility of a paradox, in which conflicting masses are to be assigned to a particle, because a specific mass is determined by the physical situation of the particle.

7. THE RELATIVITY OF TIME.

In the preceding four sections I have described the application of the Lorentz time transformation equation to the time characteristics of four different processes. In two of these, namely the variations of mass-energy and of meson decay time with velocity, the process is at a fixed point in the moving coordinate system, S', and we therefore take $\delta x'=0$ in the Lorentz time equation, (1), section 4. In gaining the equation for de Broglie waves we also take the process to be at rest in S', but the matter-wave process is extended over a distance δx as observed in the stationary coordinate system, S, with $\delta t'$ equal to zero in S' (Eq. (1), sec. 5). In the Doppler effect (sec. 3), the process considered is moving with a speed c in both S and S', and has an extension $c\delta t$ or $c\delta t'$. We have therefore seen the effect of relativistic time

[11]R. Schlegel, *Amer. Journ. of Physics, 22,* 77 (1954).

transformations on processes which are effectively at a point in a moving system and on processes which are spatially extended.

It is commonly said that one of the great philosophical consequences of relativity theory is the explicit rejection of an absolute time in the physical world. We see this rejection clearly justified in the effect of relative velocity on meson decay times and on the fundamental intrinsic frequency (and hence mass-energy) of a mass particle. A process, like the meson decay, which requires a given time in one coordinate system is found to occur in a different time when observed from a different and relatively moving system. Or, likewise, a fundamental frequency in S' is found to be something else in S. There is no measure of time other than that given by physical processes. If these processes are affected by relative motion with respect to an observer, then also "time" is affected by motion, and there is no universal, absolute time scale for *all* processes.

Relativity of simultaneity is a corollary of the relativity of time interval, as developed in the special theory of relativity. If there were an absolute time scale, there would be no question about whether or not any two events occur at the same time. We see however from the Lorentz transformations that the time interval between two events will vary with the coordinate system. Thus, if the time interval $\delta t'$ is zero between two events observed from S', Eq. (2) of section 5 tells us that the events will be separated by a time interval $\delta t = v\delta x/c^2$ for an S observer. Only if the two events occur at a single point, such that $\delta x'$ and therefore also δx are zero (or are separated by a space interval along which there is no relative velocity of observers, as, e.g., with a separation in the y-coordinate and relative motion along the x-coordinate) will there be the possibility of the events' appearing as simultaneous to all observers.[12]

De Broglie waves are a physical exemplification of the effect of relative velocity on simultaneity, as has been discussed in section 5. In the system in which a mass particle is at rest, the parts of a matter wave are at the same time over all space points, and there is therefore no

[12]It has often been pointed out in discussions of relativity theory that if propagation of signals with infinite velocity were possible, we could maintain absolute simultaneity even with the difference in clock rates that results from relative motion. The time of two clocks, A and B, spatially separated, could then be instantly observed from all coordinate systems, and there would be no ambiguity about whether or not two clocks read the *same* time. An absolute reference frame would in effect then be determined, since the coordinate system in which two spatially separated clocks had no time lag (still assuming that relative motion gave rise to such a lag) would be a natural preferred reference frame. However, in the application of the time transformation to a specific physical situation, like the waves of a mass particle, the problem of comparisons between readings of a given pair of clocks, from two or more different coordinate systems, cannot even arise. The particle waves themselves constitute the "clock," and a given particle in a given situation can be observed from one coordinate system only, i.e., from the coordinate system of the physical apparatus with which it is undergoing interaction. The temporal effects of relative velocity on the wave "clock" are immediate and unique, for a given relative velocity.

time or phase difference at different space points. To a moving observer there is a time difference and hence a phase difference associated with the matter wave at different points; for that observer there is therefore a finite and observable wave-length. In the Doppler effect equation use is also made of the time interval associated with a moving space interval. The effect is however perhaps not properly an illustration of relativity of simultaneity, since in no physically realizable coordinate system would δt become zero in some other system; that is, if in one system $\nu' \neq 0$ for radiation from a given source, we cannot go to another system for which $\nu = 0$ except by moving with a velocity $v = c$, and to give a material system such a relative velocity would require infinite energy.

There are many uses of the relativistic time transformations in physics other than the four processes that we have discussed. In the theory of interactions of elementary particles with each other and with electromagnetic and mesonic fields, there is wide need for the Lorentz transformation equations; the successes of this physical theory are further confirmation of the validity of the special theory of relativity.

We have emphasized in our discussions of the four relativistic time processes that the time transformation is applied to a specific physical interaction. Thus, a given meson decays into the counters of one given coordinate system, the mass or wave-length of an electron is measured at any one time in the apparatus of a single coordinate system, and a photon of electromagnetic radiation interacts at any instance with a single observer or recording apparatus. There is also, however, a large body of doctrine and discussion in which the time transformations are applied to processes or systems which are complex, in the sense that they obviously involve many internal processes which are not themselves directly involved in physical interactions to which the Lorentz transformations need be applied. A large-scale clock, or a biological organism, are examples of these complex systems. The processes of these systems are "public," in that they may be observed simultaneously from many coordinate systems, with no effect on the complex processes. The extension of the relativistic time transformation to these processes is altogether hypothetical, since there is no experimental evidence regarding relativistic effects on literal clocks or other such public systems. Nonetheless, the possible validity of such an extension is of great physical and philosophical importance, and we shall consider the problem of these applications of the Lorentz transformations in the next two chapters.

General References for Chapter VI.

The Theory of Relativity, by C. Møller, Oxford University Press, 1952, and *Introduction to the Theory of Relativity,* by P. G. Bergmann, Prentice-Hall, New York, 1947, are two standard textbooks in relativity theory. Four other notable books in English on relativity are: Hermann Weyl, *Space-Time-Matter,* trans. H. L. Brose, Dutton, New York, 1921; A. S. Eddington, *Mathematical Theory of Relativity,* 2nd ed., Cambridge University Press, 1924; R. C. Tolman, *Relativity, Thermodynamics, and Cosmology,* Oxford University Press, 1934; J. L. Synge, *Relativity: the special theory,* North-Holland Pub. Co., Amsterdam, 1956. Albert Einstein wrote a popular exposition, *Relativity: The Special and General Theory,* trans. R. W. Lawson, Holt, New York, 1922. His *Meaning of Relativity,* Princeton University Press, 5th ed., 1955, is to be recommended for an advanced rather than a beginning student of relativity theory. A good quasi-popular account is *Einstein's Theory of Relativity* by Max Born, revised edition, Dover, New York, 1962.

Chapter VII

Clocks and Special Relativity Theory

1. THE TRANSFORMATION OF CLOCK TIMES.

In his first paper on the special theory of relativity, Einstein wrote: "If at the points A and B of (system) K there are stationary clocks which, viewed in the stationary system, are synchronous; and if the clock A is moved with the velocity v along the line AB to B, then on its arrival at B the two clocks no longer synchronize, but the clock moved from A to B lags behind the other which has remained at B by $\frac{1}{2}tv^2/c^2$ (up to magnitudes of fourth and higher order), t being the time occupied in the journey from A to B."[1] The doctrine that a clock will undergo a change in time rate as a result of motion with respect to an observer was thus included in the foundations of special relativity theory, and the successes of the theory have been sufficiently impressive that in recent decades physicists have in general shown little inclination to question the assertion.

We can readily derive the equation for the rate change of a clock, by the use of the Lorentz transformation equation,

$$\delta t = \frac{\delta t' - (v/c^2)\,\delta x'}{\sqrt{1-\beta^2}}, \tag{1}$$

where as usual, $\beta = v/c$.

We assume that a clock is at rest in the system S', which, as before, moves with a uniform velocity $v = v_x$ with respect to S. This clock may be compared with clocks of altogether similar construction which are at rest in S. The clock at rest in S' is at a fixed x' point; so $\delta x' = 0$, and we may write at once from (1),

$$\delta t = \delta t' / \sqrt{1-\beta^2}. \tag{2}$$

Eq. (2) is identical with Eq. (2) of section 4, in the previous chapter, which was applied to the mu-meson decay time as observed from S or

[1]See Reference 1, Chap. VI. The sentence quoted is on p. 49, English trans., *The Principle of Relativity,* Dover, New York.

S'. We now apply the equation to a clock, and state that δt is the time interval which is presented by a clock at rest in S; that is, any observer who looks at an S clock will read the interval δt. Likewise, $\delta t'$ is the time interval which any observer will find for the clock at rest in S'. It follows from (2) that

$$\delta t' = \sqrt{1-\beta^2}\,\delta t, \tag{3}$$

or, that the clock in S' runs slower than the clocks at rest in S. This derivation of the difference in clock rates is essentially the same as that given by Einstein in the aforementioned paper. An expansion of the factor $\sqrt{1-\beta^2}$ in (3) gives $\delta t' = \delta t(1-\frac{1}{2}\beta^2 - \frac{1}{8}\beta^4 \ldots\ldots)$, so that $\delta t - \delta t' = \frac{1}{2}(v^2/c^2)\,\delta t$, neglecting powers of the fourth and higher magnitude, as stated in the quotation.

It is important to note that in applying Eq. (2) to a clock we must give an interpretation to the variables of the equation that is different from the interpretation used in the application to the various physical processes that were discussed in the preceding chapter. The frequency of a photon or particle, for example, was taken as being temporally characterized by the variable t for an observer in S and by t' for the S' observer. Likewise, the meson decay time is δt or $\delta t'$, for S or S' observers respectively. But a clock is a "public" process which can be simultaneously observed from any number of different coordinate systems, and therefore cannot present more than one time interval. Hence, as has been noted, the clocks at rest in S' are taken to *show* the time interval $\delta t'$ to *any* observer. The clocks at rest in S, likewise, show the interval δt. The readings of the S' clocks are then the same for S' and S observers, but, because all S' clocks are slower than all S clocks, the S observer will find S' clocks to have a slower rate than his own rest clocks. Another way of seeing why we must interpret Eq. (2) as we do for clocks is to realize that a moving clock is both a process whose time rate may be considered and *in itself* a measuring apparatus. The energy-frequency of an electron, for example, is determined with reference to rest measuring apparatus in the observer's system. Likewise, the time between two readings of the moving clock may be determined by the S observer by reference to his rest clock, and he states that time interval as being δt. But *also* (in contrast to other relativistic processes that we have considered) the clock can itself give its own time reading $\delta t'$ for the interval; and this is the interval in the rest system S' of the moving clock. The electron frequency quantities ν and ν' (or m and m'), for example, may be for any two relatively moving systems S and S' with respect to whose apparatus the electron mass is determined. For the clock, its reading for an interval, $\delta t'$, must *always* be that interval in the clock's rest system S', and δt may be the interval as judged by clocks in any relatively moving system S.

If we write the Lorentz time transformation equation in the form,

$$t' = \frac{t - vx/c^2}{\sqrt{1-\beta^2}}, \tag{4}$$

we see that the equation, applied to clocks, not only leads to the rate difference expressed in (3), but also asserts that there is a difference with position in clock times, as the clocks are observed from a moving system. Let us suppose that rest clocks are placed along the (+) and (−) x axes in S, and likewise along the x' axes in S'. All the clocks are of similar physical construction, and the clocks in any one system are synchronized by means of light signals sent from one clock to the other. This means that if a signal is sent from any clock C_1 in, say, the S system, at time t_1, to another S clock, C_2, the signal is received at C_2 at a time $t_1 + d/c$, where d is the distance between C_1 and C_2 and c is the speed of light propagation. Further, for simplicity let us assume that the S clock at $x=0$ reads $t=0$, at the instant which we now consider. By Eq. (4), $t'=0$ also, so the clock at the point in S' opposite $x=0$ will also read 0 at this instant. But a clock in S' at a point which has the position coordinate x, in the S system, will have a reading $\frac{-vx/c^2}{\sqrt{1-\beta^2}}$. In other words, the S observer, who is satisfied that all of his clocks read $t=0$ at a given instant, will find that at this instant the various clocks in the S' system have readings which are different from 0, excepting for the S' clock at $x=0$. We imagine that there is an observer stationed with each S clock, at various x points in S. These observers, at the instant that their clocks read $t=0$, note the times of the S' clocks, moving by the various x points with velocity v. The observers find readings that vary from $\frac{-vx_i/c^2}{\sqrt{1-\beta^2}}$, for the S' clock at x_i, the point of greatest $+x$ magnitude, to $\frac{+vx_f/c^2}{\sqrt{1-\beta^2}}$, for the S' clock at x_f, the point of greatest $-x$ value.

The S' observer, however, would not think that his clocks are out of synchronization. Rather, he would think that the S observers did not read his clocks at a common instant, but instead, made their first reading of his clocks when one of these clocks was at x_i, at time $\frac{-x_i v/c^2}{\sqrt{1-\beta^2}}$, and then made other readings at successively later times, finally ending with a reading of his clock as it passed point $-x_f$, at (S') time $\frac{x_f v/c^2}{\sqrt{1-\beta^2}}$.

The Lorentz transformation equation,

$$t=\frac{t'+vx'/c^2}{\sqrt{1-\beta^2}}, \tag{5}$$

would, in contrast to (4), give the values of the S clocks, as they passed various S' points x', and as observed from S' at a common time $t'=0$ on the S' clocks. Eqs. (4) and (5) express the special relativity theory result that events (in this case clock readings) which are simultaneous in one system will not be so if viewed from another system. The de Broglie equation, discussed in section 5 of the preceding chapter, is, we saw, a consequence of this dependence of simultaneity on coordinate system.

We recall that the Lorentz transformations are based on the assumption that all observers will find the same value c for the speed of light. If we ensure this invariance of c by requiring that clocks and measuring rods in different systems have different properties, it is readily seen why there must be a difference in clock values with position, as has just been discussed. If we refer to Fig. 5 of section 1, Chapter VI (p. 83), we note that both the observer in S and the one in S' must find the spherical light signal from O, O' to be spreading uniformly. At an instant t when the O observer finds the wave front of the signal to be a perfect sphere, the O' observer finds the wave front to be distorted, so that it is much closer to him in the $+x'$ direction than in the $-x'$ direction. However, this "instant" for O is not a single instant for O'; rather, his clock opposite point x reads $t'=\frac{t-vx/c^2}{\sqrt{1-\beta^2}}$, when the wave is at x, and his clock opposite $-x$ reads a later time, $t'=\frac{t+vx/c^2}{\sqrt{1-\beta^2}}$. He expects, then, that the distance to the wave should be much farther along his $-x'$ than along the $+x'$ axis, since by his clocks the wave is at the $-x'$ point at a time $\frac{2vx/c^2}{\sqrt{1-\beta^2}}$ later than when it is at the x' point. Conversely, at an instant t' when the wave is a perfect sphere to the S' observer at O', the wave positions correspond to different times t for the S observer at O.

2. THE CLOCK PARADOX.

In setting aside the ether as the reference system for the state of motion or rest of a body, the authors of special-relativity theory implicitly introduced a symmetry into the effects which arise as a result of relative motion. Observers in S may find relativistic changes in processes at rest in S', but then also, observers in S' should find those same effects for similar processes at rest in S; an observer in either system may justifiably consider his system to be at rest and the other to be in motion. It however then follows that if two observers with clocks are moving relative to each other, each should expect to find the other's clocks to be running slower than his own. This situation is the essential element of the famous clock paradox of relativity theory. The paradox has been discussed in many books and journal articles, and these discussions present a variety of proposals for solution and also many criticisms and counter-proposals. Currently the problem is attracting heightened interest because of the discussions of possible high-speed rocket travel: a human being, or other biological organism, is a kind of clock, and the question arises, would biological aging be delayed as a result of such rocket travel?

The clock problem was first stated by P. Langevin[2] in 1911, and already in his original statement the presumed change of time rate was applied to human beings. In later statements, the people involved have commonly been referred to as twin brothers, Peter and Paul. Peter remains on the earth, while Paul is fired into space in a projectile, with a speed v such that the factor $\sqrt{1-\beta^2}$ is 0.01. As a result of an interaction with a star, Paul's projectile is eventually reversed, and he comes back to earth at the same high speed at which he left. Suppose that he has been gone two years, by his clock; then, according to clocks on the earth, the journey will have taken two hundred years. Paul will have aged only slightly, while Peter will have long since died and will be represented by descendants several generations removed. However, we might equally well take Paul to have been at rest, and Peter to be the moving twin. Then, it is Peter who should age only two years while two hundred years have passed in Paul's reference system. We have the paradox that each twin may take the other to be moving, and therefore to be aging at a rate slower than his own.

The fact that a variation from uniform motion is required to bring one clock (or twin) back to the other has been emphasized in many proposed solutions of the paradox, and in his first statement of the problem, Langevin avoided paradox on the basis of this dissymmetry of the motions. He gave no calculations, but pointed out that if the earth and traveller were in communication by means of electromagnetic waves, the traveller would on turning back toward the earth at once detect a shift to shorter wave-lengths in the waves that he was receiving from the earth. The earth observer, in contrast, would not find a wave-length change in waves received from the projectile until a time L/c after the projectile's turning, where L was the distance of the projectile from the earth when it turned.

[2] *Scientia, 10,* 31, 231 (1911). L. Lange, in an article on "The Clock Paradox in the Theory of Relativity," *Amer. Math. Mon.* XXXIV, 22 (1927), gives a concise account of the paradox and of various proposals for its solution.

Langevin's criterion for distinguishing which of two relatively moving bodies changes its state of motion is a valid one; we might think that radio messages from each body set up an electromagnetic field about that body, and the body which changes its motion at once detects a wave-length change in the "message field" of the other body. However, the question of a criterion for motion is in any case of little relevance, since in accordance with the basic principle of relativity, any effects of motion should be reciprocal between two relatively moving systems. It is easy, also, to formulate the paradox in a manner such that there is complete symmetry between the moving systems. We think of two rockets, R_S and R_K, of identical physical construction, which move past each other at a point P. This point is in a reference frame such that R_S moves away from P with a velocity $+w$ and R_K moves from P in the opposite direction, with velocity $-w$. The rocketeers in R_S and R_K observe each other's clocks at the moment of passing at P, and each finds his and the other rocket's clocks to read zero. After a time, properly chosen by each rocketeer, each rocket spews out an identical amount of hot gas, causing it to reverse its velocity, so that at some time after the first coincidence the observer at P finds the two rockets again to be passing at his point. Each rocket underwent the same period of acceleration at its turning point, so that any effect of acceleration on clocks would be the same on each rocket. Further, that effect can be made as small as desired compared with the effect of uniform motion on the clocks, by extending the period of uniform motion for as long as is desired. The turning effect would be the same however long the trip. The rocketeers in R_S would expect, then, that at the second passing at P the clocks in R_K would be slower than theirs by an amount $t(1-\sqrt{1-\beta^2})$, where v is the relative velocity between the rockets and t the reading of the R_S clock. The R_K rocketeers would have a similar expectation for the R_S clocks. But clearly, both cannot be correct.

3. PROPOSED SOLUTIONS.

We shall consider the manner in which one can suggest a fallacy in the "clock paradox" as exemplified in the rocket clocks of the preceding section. Before doing this we shall first, however, treat a simpler situation, thereby seeing a principle that will be helpful in our treatment of the rocket clocks.

We consider a motion in which one clock moves with respect to the clocks of a stationary coordinate system; it might seem that even with no "turning" and second coincidence of a pair of clocks we reach a paradox, since the moving clock should be running slower than the stationary ones that it passes, and, contrariwise, the observer moving with the clock should find the stationary clocks running slower than his own clock. A simple calculation shows both of these expectations to be borne out, and yet that nevertheless there is no paradox.

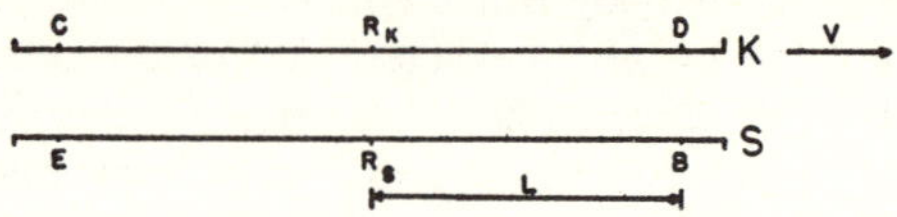

Fig. 8
A clock, R_K, moves with system K at a velocity v with respect to S. The moving clock R_K is compared with clocks that are fixed at various points in S.

We denote the stationary system as S (Fig. 8), and the moving clock, R_K, is at rest in the system K, which has a velocity $v=v_x$ with respect to S. In the system S there is a fixed clock R_S, and at the instant of R_K's passing R_S—the (R_K, R_S) spatial coincidence—both R_K and R_S read zero. Also, at this instant other fixed clocks in the S system, and notably the clock at B, l distance from R_S, read zero to the S observer. These S clocks, all stationary with respect to each other, have been synchronized, as by passage of light signals between them, so that a signal is received at any one clock at a time d/c later than the time at which the signal was sent from the clock at distance d. The S observers, then, find the clock R_K to pass R_S at $t_{R_S}=0$ and to reach the clock B at $t_B=l/v=\tau$. When R_K is opposite B, the reading on R_K is given by the Lorentz transformation equation, $t_K=\dfrac{t_S-vx_S/c^2}{\sqrt{1-\beta^2}}$. In this case, $t_S=t_{R_S}=\tau$, $x_S=l$, $t_K=t_{R_K}$, and we have $t_{R_K}=\dfrac{\tau-vl/c^2}{\sqrt{1-\beta^2}}$, or, since $l=v\tau$, $t_{R_K}=\tau\sqrt{1-\beta^2}$. This value for R_K is in agreement with the expectation of S observers that the moving K clocks run slow, since for S the time of movement of R_K from R_S to B is τ.

Now, from the point of view of K observers, the motion just discussed is one in which the S clock at B moves from a point D, in K, to a point opposite R_K. At the (R_S, R_K) coincidence, although R_K reads $t_K=0$, the K clock at D will not read $t_D=0$; for, by application of $t_K=\dfrac{t_S-vx_S/c^2}{\sqrt{1-\beta^2}}$, we find $t_D=\dfrac{t_B-vx_B/c^2}{\sqrt{1-\beta^2}}=\dfrac{o-vl/c^2}{\sqrt{1-\beta^2}}=\dfrac{-\tau v^2/c^2}{\sqrt{1-\beta^2}}$. That is, the spatial coincidences (R_K, R_S) and (D, B), which are both at $t=0$ to the S observers, do not occur at the same instant to the K observers. When R_K and B are at the same point, t_{R_K}, as we have seen in the preceding paragraph, is $\tau\sqrt{1-\beta^2}$. The time for the motion of B from D to R_K, as observed from K, is given then by $t_{R_K}-t_D=\tau\sqrt{1-\beta^2}+\dfrac{\tau\beta^2}{\sqrt{1-\beta^2}}=\tau/\sqrt{1-\beta^2}$. The S clock at B reads only τ when

B reaches R_K, and this is as K observers expect, since for them it is the S clock which runs slower than their own, by a factor $\sqrt{1-\beta^2}$. We have no paradox, and observers in each system find clocks in the other system to run slower than their own. The essential principle used in coming to this conclusion is the "relativity of simultaneity." It is because the (D, B) coincidence appears simultaneous with the (R_K, R_S) coincidence to S observers, but appears to occur earlier than the (R_K, R_S) coincidence to K observers, that each set of observers can conclude that the other's clocks run slow. There is no conceptual difficulty about the difference in judgment with respect to the simultaneity; the K observer, by his synchronized K clocks, merely notes that the clock B passed his clock D at a time $\frac{\tau\beta^2}{\sqrt{1-\beta^2}}$ before clock R_S passed his clock R_K.

We now consider the two rockets. We may describe the situation by reference to Fig. 9. To an observer at P, rockets R_S and R_K, after a

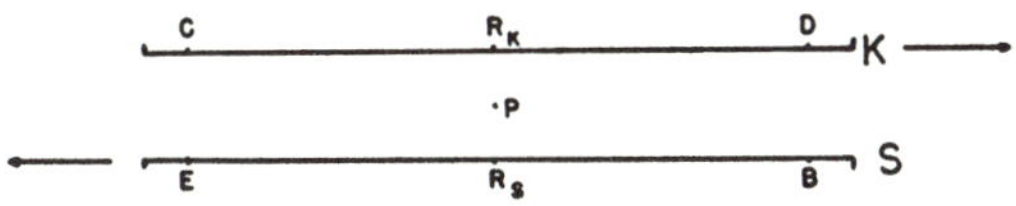

Fig. 9
Rockets R_K and R_S are in spatial coincidence as they move past point P, with respect to which they have equal speeds in opposite directions. R_K turns around at point B in the system S of R_S; R_S turns at point C of the system K of R_K. Both rockets return for a second coincidence at P.

coincidence at P, move away with equal speeds in opposite directions along the x axis. Rocket R_K turns around when it is opposite a point B in a hypothetical coordinate system S attached to the rocket R_S, and rocket R_S turns after it has gone to the point C in the coordinate system of rocket R_K. R_S and R_K are at the origins of their coordinate systems. After turning, each rocket returns to P, reaching the point for a second (R_K, R_S) coincidence. For the R_S observer, the motion may be described as one in which there is a relative velocity v between the two rockets. R_S observers find that their rocket turns when at the point C of the R_K system, and that R_K turns at the point B of their system. Similarly, R_K observers find a relative velocity v between the rockets, with turnings at the points B and C. For both rockets, the motion of interest is between the first and second spatial coincidences (R_K, R_S).

The observers in R_S set the clock in their rocket at $t=0$ at the (R_K, R_S) coincidence, and also for them the S system clocks at E and B are at zero at the coincidence. R_K is observed to reach its turning point B at a time $l/v=\tau$, where l is the distance of B from R_S. That is, clocks R_S and B read τ when R_K is at B. R_S is at C at this same time, τ, so that the R_S observers consider R_K and R_S both to be turning at the time τ. The clock of R_K, by the R_S calculation, will read $t_{R_K}=\frac{t_B-vx_B/c^2}{\sqrt{1-\beta^2}}$ when R_K is at B. Or, since $t_B=\tau$ and $x_B=v\tau$, $t_{R_K}=\tau\sqrt{1-\beta^2}$; the R_K clock will have run slower by the usual relativistic factor $\sqrt{1-\beta^2}$. The "return" journey to the second (R_K, R_S) spatial coincidence will again require τ seconds by the R_S clocks, so that R_S reads 2τ at that coincidence; the journey of R_K back from B to R_S will require $\tau\sqrt{1-\beta^2}$, as calculated by R_S, so that R_K will read $2\tau\sqrt{1-\beta^2}$ at the second coincidence.

For the R_K observers, their first coincidence with R_S is at $t_{R_K}=0$. They then observe R_S to move to their point C before turning, and they move to the point B, where they turn. As has just been calculated, $t_{R_K}=\tau\sqrt{1-\beta^2}$ at the (R_K, B) coincidence; so for the R_K rocketeer, his journey away from P is through a distance $v\tau\sqrt{1-\beta^2}$ rather than $v\tau$ (as for R_S). Further, in considering R_S, the R_K observer finds a different result than for his own trip. R_S moves from R_K to C, and the time t_C at which R_S reaches C is given by the Lorentz transformation $t_C=\frac{t_S-vx_S/c^2}{\sqrt{1-\beta^2}}$. Since t_S is $t_{R_S}=\tau$, and $x_S=0$, $t_C=\tau/\sqrt{1-\beta^2}$. Hence, for K the "outward" journey of R_S runs from the time $t_{R_K}=0$ of the first (R_K, R_S) coincidence to t_C, or for a time $\tau/\sqrt{1-\beta^2}$. So, for the R_K observers the two rocket trips are not identical: R_K goes out for a time $\tau\sqrt{1-\beta^2}$, and turns back to P; R_S goes out for a longer time, $\tau/\sqrt{1-\beta^2}$, and the instants of turning are not simultaneous. The difference in the outward journeys is $\tau/\sqrt{1-\beta^2}-\tau\sqrt{1-\beta^2}$, which is equal to $\beta^2\tau/\sqrt{1-\beta^2}$. The R_K observer therefore calculates that, since his total trip requires a time $2\tau\sqrt{1-\beta^2}$, the R_S total trip takes a time $2\tau\sqrt{1-\beta^2}+2\beta^2\tau/\sqrt{1-\beta^2}$. Also, however, R_K expects the R_S clocks to run slower than his by a factor $\sqrt{1-\beta^2}$, so that the R_S clock at the second (R_K, R_S) coincidence should read $\{2\tau\sqrt{1-\beta^2}+2\beta^2\tau/\sqrt{1-\beta^2}\}\sqrt{1-\beta^2}$. Algebraic reduction shows this product to be 2τ, which is just what the R_S rocketeer's clock was originally assumed by R_S calculations to read at the return to P.

We find, then, that we do not come to paradox. The observer in each rocket, on making allowance for the other rocket's clocks running slower than his own, calculates the same set of values for the various clocks involved. Paradox is avoided, because the two rocket trips are not symmetrical for the R_K observer, although they are so for the R_S observer. There is also the asymmetrical result, that for R_S the total elapsed time between the two (R_S, R_K) coincidences is 2τ, while for R_K the time is $2\tau\sqrt{1-\beta^2}$; this is so, even though R_K takes into account the fact that R_S clocks run slow by the factor $\sqrt{1-\beta^2}$. Since both R_K and R_S observers agree that the R_K time between the coincidences is less than R_S time, the R_K observers would have to conclude that their time is genuinely running slower than is R_S time.

Some students of the clock paradox have asserted, on the basis of the agreement that one can find for the calculated clock times in two relatively moving coordinate systems, that the problem of the paradox is altogether illusory.[3] A statement of the paradox in the form that we have presented it shows, however, that it cannot be readily dismissed. We did find that on the assumption of a clock setting in the R_S coordinate system such that the R_K and R_S motions are symmetrical from the R_S point of view, that we came to no paradox. However, we might equally well have asserted that the motion of the two rockets is symmetrical in time from the R_K point of view, and then it would be the R_S observers who find that they travel for a shorter time than does the R_K rocket, and at the second (R_K, R_S) coincidence it would be the R_K clock which reads 2τ, and the R_S clock $2\tau\sqrt{1-\beta^2}$. Or, finally, we might have considered the motion only from the point of view of the P observer. For him each rocket travels at a relative speed w, and if his clock reads zero at the first (R_S, R_K) coincidence and t at the second, he would expect the clocks of each rocket to read $t\sqrt{1-w^2/c^2}$ on its return to P. In a word, the choice of original coordinate system will determine whether R_K and R_S clocks read the same at the second (R_K, R_S) coincidence, or whether it is R_S or R_K that gives a greater reading than the other system.

What is shown by calculations with the Lorentz transformations in situations of relatively moving clocks is that given a set of original conditions, the transformations will give us a set of consistent results for all the clocks involved. That is, if we start with the settings of the synchronized clocks in one system, and determine times of all other clocks by transformations from that system, we do not come to any paradox. (We would hardly expect to do so, since the Lorentz transformations form a self-consistent group.) The problem that is not solved, however, is what system shall be taken as the "base system." The "clock paradox" is not, then, strictly a paradox. Rather, it might be called the "clock ambiguity," because the essence of the problem is: With respect to what coordinate system shall we find the equations that will describe the actual, physical behavior of clocks? The consistency of the Lorentz transformation equations is such that there will

[3]W. H. McCrea, *Nature, 167,* 680 (1951), *177,* 782 (1956), *178,* 680 (1956); A. Grünbaum, *Philos. of Science, 21,* 249 (1954); B. Leaf, *idem., 22,* 45 (1955); E. M. McMillan, *Science, 126,* 381 (1957), *127,* 161 (1958). J. W. Campbell gave a complete solution, on the assumption that acceleration is with respect to the universe, and one clock is accelerated and the other is not: *Phil. Mag., 15,* 48 (1933), *16,* 529 (1933).

be no paradox, whatever be the coordinate system which we do take as our base system.

It may be worthwhile to give some further discussion of the problem of ambiguity of the base system. Suppose our rockets are considered by the stationary "middle" observer at *P*. Let us say that he finds them to pass point *P* at $t_P=0$, to go equal distances before "turning," to turn at identical times by his clock, and to return to his position at the same instant of time. For this observer, the clocks of both rockets R_K and R_S will run slower than his own, and application of the Lorentz equations will give R_K and R_S clock readings that are consistent with the results predicted by the rest observer at *P*. Further, the R_K and R_S readings will show that the turnings of rockets R_S and R_K, at *C* and *B* respectively, are not simultaneous to R_S and R_K observers, even though they are simultaneous to the *P* observer. Now, we have shown in the preceding calculations that we may also take it that the rocket R_S (or R_K) observer finds the two rocket turnings to be simultaneous. We then get another and different (but also internally consistent) set of readings for the clocks at *P* and in rockets *S* and *K*. It might be argued that the *P* observer is the "right" observer, because he is "stationary"; but then we would in fact be appealing to a preferred coordinate system, and the principle of relativity forbids our doing so. The observer in the *S* rocket, for example, also has the privilege of considering his system to be the rest system, and of saying that it will be by his clocks that the rockets R_S and R_K turn simultaneously, if they are to return simultaneously to *P*, and that the clocks at *P* and in rocket *K* will be slower than his at the second (R_K, R_S) coincidence.

We might say, we settle the matter by setting up the rocket situation as rest laboratory observers, associated with the point *P*, taken as a rest laboratory mid-point. We arrange the rockets to "turn" simultaneously and symmetrically at points where a pair of laboratory clocks, equi-distant from *P*, have the same readings. Our base system, then, is clearly the laboratory system. But we thereby "settle" the matter only if we assume that the rocket clocks "know" that the laboratory system is the base reference system for clock times and behave accordingly; that is,.if the rockets behave as if they were moving with respect to the laboratory system, with their clocks related to the rest clock synchronization of *that* system by the Lorentz transformation, and not *vice versa*. In a word, the matter is settled only if the laboratory system is functioning as a preferred rest system.

In physical fact, the R_K and R_S rocket operators would find (say by trial and error adjustment) that each must turn at some reading of his clock, in order to meet again with the other rocket at *P*. If each rocketeer found that he turned at the same time, by his clock, and each had the same reading, slower than *P*'s clock reading, at the second meeting at P, then the laboratory system would be functioning as a preferred coordinate system in the relative motion of the clocks. But, it might be found, as we showed in the calculations, that the turnings occurred so that they were simultaneous only in the S rocket system; and then, in accordance with the Lorentz transformations, we would expect the R_K and P clocks to be slower than the R_S clock at the second meeting at P. In this case, the R_S system would be functioning as a preferred coordinate system. *A priori,* the special relativity theory gives us no criterion of preference between the system of R_K, R_S, or P, with respect to the question of whose clocks shall be slow at the second rocket coincidence at P.

The original "twin paradox" problem entails the same ambiguity as does the rocket problem. We get no paradox, taking the system of one or the other twin as our base system, and we find one or the other twin to be younger on return from his journey in space. The question is, whose system is the base system of synchronized clocks from which we calculate times in the two coordinate systems; or, *which* is the twin that will be younger? Most physicists to-day seem to be willing to accept as an adequate solution to the paradox the completely consistent solution of the problem that we gain by taking the rest system as the base system. We can also, to be sure, get a consistent, and different, solution if we use the moving system as a base system. But, the latter system is not at all times a single inertial system, since it must undergo acceleration at its turning point, and it is tacitly argued that therefore the moving system is not a proper "base system." In a word, we introduce the asymmetry that will resolve the paradox by asserting that the rest system clocks, which *at all times* are in a single inertial system, constitute the proper base system. We make our calculations initially from that system, and when we calculate the times from the moving system we do so on the assumption that the clocks in that system are related by the Lorentz transformations to the clock or clocks of the rest system.

It is not an easy matter to decide whether or not the solution to the paradox that accepts the preferred role of the rest coordinate system is an entirely legitimate one. It is indeed true, that if any basis at all can be granted for the distinguishing of one system as a "preferred" system, in a clock paradox problem, then the resolution of the paradox is easily made, as a consequence of the internal consistency of any group of Lorentz transformations. It is worthy of note that a recent thorough discussion of the twin paradox came to the point—similar to the one that we have reached with respect to our rocket problem—that acceptance of a preferred coordinate system is necessary in order to determine which twin is to be the younger on return. In an extended series of papers[4] in *Nature,* in which various authors discuss the paradox, the defenders of the traditional "moving clocks will run slower" view have argued that the application of the Lorentz time transformation to clocks or biological systems does not in itself lead to paradox, but seem also to have conceded that some appeal to a coordinate system is required to define which of two clocks in a twin paradox situation will be slower than the other. In the fifth of the *Nature* papers cited, F. S. Crawford, Jr., who argues for the slowing-of-clocks view, states that the "accepted" solution of the twin paradox (a solution in which one twin is found unambiguously to age more rapidly than the other) does rest on reference to a "third body," as "the matter in the remainder of the universe" or the frame of "the distant galaxies."[5]

The "clock paradox" is, then, in a sense resolved within the framework of special relativity theory, in that we find a consistent solution to any clock problem, assuming some initial set of synchronized clocks. Any legitimate preference among coordinate systems will give a basis for resolution of the paradox: we may argue that since the Lorentz transformations are applicable between inertial coordinate systems, a system which at all times is an inertial system has a preferred status with respect to the transformations; or we may argue that the total matter of the universe determines a preferred coordinate system. If we do not accept and justify some one preferred coordinate system, we will find that we can calculate different (although internally consistent) sets of clock values from the different coordinate systems of a clock problem. We shall discuss this matter again in Chapter VIII. A discussion of a possible solution of the clock paradox in terms of general relativity theory is, however, now in order.

4. THE GENERAL RELATIVITY SOLUTION.

A solution of the clock paradox using the results of general relativity theory was suggested by A. Einstein in 1918.[6] A problem like that of the twin brothers is considered, with one staying at rest and the other being accelerated to a high velocity, travelling for a relatively long period at uniform velocity, then turning around and eventually rejoin-

[4] *Nature, 177,* 782 (1956); *178,* 680 (1956); *179,* 36, 865, 909, 977, 1071, 1242 (1957). The major protagonists in this series are W. H. McCrea and Herbert Dingle. Professor Dingle has consistently (and in general I think justifiably) criticized the asymmetry which is introduced in the usual solutions of the paradox. His own proposed solution, however, does not give due account to the literal relativistic transformations of time rates that are observed for physical processes. Further discussions involving Max Born, H. Dingle, W. H. McCrea, and others are in *Nature, 195,* 985 (1962); *197,* 1287 (1963); *216,* 119 (1967); *217,* 17 (1968).

[5] See also G. Builder, *Austral. Journ. of Physics, 3,* 279 (1958), who argues for an aether as a preferred frame in order to solve the clock paradox.

[6] *Naturwissenschaften, 6,* 697 (1918). A development in complete detail is given by C. Moller, *Dan. Mat. Fys. Medd., 20,* No. 19 (1943). The solution is lucidly presented by R. C. Tolman, *Relativity, Thermodynamics, and Cosmology,* Oxford, 1934, sec. 79 and also by L. Lange, *op. cit.,* p. 29.

ing his brother. The accelerations required during the journey of the travelling brother are taken into account, and their effects on time rate are found to remove the paradoxical situation of each brother's expecting the other to age at a slower rate than his own.

We consider A to be the observer who stays on earth, while B moves off in a rocket at a velocity v, and after a long period of travel returns to A. If t_A is the time between B's leaving and returning, by A's clock, the Lorentz transformation indicates that for B the total time of travel t_B should be related to t_A by the equation $t_A=t_B/\sqrt{1-\beta^2}$ or $t_B=t_A\sqrt{1-\beta^2}$ so that B should have aged less than A by a factor $\sqrt{1-\beta^2}$ during his trip. To a first approximation, we may write $t_A=t_B/\sqrt{1-\beta^2}$ as $t_A=t_B(1+\frac{1}{2}\beta^2+\ldots.)$. There is clearly a lack of symmetry between the physical situations of A and B during the time t_A (or t_B), in that A is not subjected to physical forces, while B is subjected to accelerations for relatively short periods of time. These periods may be indicated as τ_B', the time of B's first acceleration to a velocity v, τ_B'', the time of reversal of B's velocity, and τ_B''', the time of deceleration of B as he comes to rest with A at the end of his journey.

On the basis of the principle of relativity of motion, A may however be considered by B to be in motion, so that B finds time to move slower for A than for himself, with $t_A=t_B\sqrt{1-\beta^2}=t_B(1-\frac{1}{2}\beta^2+\ldots)$ at the end of the journey. The general relativity solution accepts this result from special relativity theory, but also considers what the effects are of hypothetical forces that act during the periods of acceleration, these forces taken to be equivalent to gravitational fields. It is a consequence of the general theory of relativity that the periods of two clocks that are in regions that have a difference in gravitational potential of ΔV will vary by the equation[7]

$$\tau_2=\tau_1(1+\Delta V/c^2) \qquad (1)$$

where the longer period τ_2 is of the clock at the lower potential. Thus, light emitted from an atom on the surface of a heavy star comes from a source that is at a lower gravitational potential than is the surface of the earth, and that light should therefore be observed on the earth as having a longer period (wave-length) than light from a similar atom on the earth.

With the point of view that B is at rest, A is now taken to be initially accelerated to a velocity v during a time τ_A', turned around during a time τ_A'', and brought to rest with B during τ_A'''. A gravitational field brings about each of these accelerations, with resulting

[7]See R. C. Tolman, preceding reference, p. 193.

effects on A and B clocks. The fields which cause the acceleration of A act also on B, but we assume that those same forces which previously accelerated B now act as counter forces which hold B at rest, so that relative to B, A is accelerated, turned about, and eventually brought to rest with respect to B. The field which is "turned on" so as originally to accelerate A, and later again "turned on" to bring A to rest, acts when A and B are in approximately the same spatial region, since A and B are at the same place at the beginning and end of the relative motion. Hence, whatever effects these gravitational fields have, the effects are the same for A and B. We may, therefore, write $\tau_A'=\tau_B'$ and $\tau_A'''=\tau_B'''$. These time intervals may therefore be neglected for both A and B, and we write $T_A=t_A+\tau_A''$ and $T_B=t_B+\tau_B''$ for the total times T_A and T_B respectively. Since $t_A=(1-\frac{1}{2}\beta^2+\ldots.)t_B$, by B's application of the special relativity time effect to A, we may write

$$T_A=(1-\frac{1}{2}\beta^2+\ldots.)t_B+\tau_A''. \tag{2}$$

The gravitational field which is "turned on" when A is farthest from B, and which causes A to turn around, is taken to be one which extends uniformly between B and A. Say A travels to a distance $h=\frac{1}{2}vt_B$ from B. In time τ_B'' there must be a velocity change of $2v$ for A, so the acceleration produced by the field is $g=2v/\tau_B''$. If this field extends from B to A, there is a gravitational potential difference of $gh=t_Bv^2/\tau_B''$ between B and A. Hence, by Eq. (1), the times τ_B'' and τ_A'' may be taken to be related by the equation

$$\tau_A''=\tau_B''(1+t_Bv^2/c^2\tau_B'')=\tau_B''+t_Bv^2/c^2. \tag{3}$$

Next, we substitute from (3) for τ_A'' in (2) and find

$$T_A=(1+\frac{1}{2}\beta^2+\ldots)t_B+\tau_B''.$$

The acceleration interval τ_B'' is, however, small compared with t_B: the period of uniform motion t_B can be considered to be as long as we wish. Hence, we may neglect τ_B'' and write

$$T_A=(1+\frac{1}{2}\beta^2+\ldots..)t_B. \tag{4}$$

But, (4) gives us the same result for T_A that we found for t_A on the original assumption that it is A who is at rest and B who moves. Hence, the argument concludes, whether we take A to be at rest and B to move, or whether we take *B* to be at rest and consider the hypothetical gravitational fields that would give *A* motion with *B* at rest, we find that the *B* clock and traveller have a slower time rate than do the *A* counterparts.

The general relativity solution has been widely presented in textbook and other discussions, as the answer to the clock paradox. And yet, the solution is a highly contrived one which can readily be criticized.[8] First, we note that the clock *B* is brought to a slower reading than that of *A*, in spite of the reciprocal nature of the special relativity slowing effect, by the effect of a hypothetical gravitational field extending linearly from *B* to *A*. And yet the existence of such a field is in no way physically justified. The forces which act on *B* to turn him around are in no way equivalent to a field extending to the position *A*. It is justifiable to argue that the accelerating force acting on *B* is equivalent to the action of a gravitational field, but this argument does not justify assumption of a field that is linearly extended from *B* to *A*. Indeed, an observer located with clock *A* would report to the observer at *B* that *A* did not undergo acceleration.

It might be thought that the general relativity approach would give a solution if the velocity change of one of the clocks is considered to be equivalent to the action of a purely local gravitational field. Such an approach cannot succeed because the effect of a local field on clock rate will be determined by the acceleration—and hence gravitational field—that is required; the special-relativity time change, in contrast, may be made arbitrarily small or large by making the period of uniform velocity small or large. Obviously, the general relativity solution will "work" only if the total distance travelled, and hence the time during which the special relativistic time change operates, is related to the gravitational field which produces the acceleration; and such a relation requires the hypothetical extended gravitational field.

Should the argument of the linear field be allowed, there is the further strong objection that when *B* is considered at rest, the assumed gravitational field that changes the motion of *A* brings the *A* clock ahead of *B*; and yet, when *A* is taken at rest, *B* undergoes a force which turns him around, but no account is taken of the effect of that force. Consistency would demand that a gravitational field be applied in that case also, rather than allowing a treatment of the motion with only special relativity theory. But if this were done, we would have reciprocity between the effects of the two assumed gravitational fields, and we would not resolve the paradox.

Finally, the general relativity approach could not resolve the clock paradox when it is expressed in the situation of the two similar rockets,

[8]L. Lange, *op. cit.*, p. 29. Also Karl Jellinek, *Weltsystem, Weltäther und die Relativitätstheorie*, Basel, 1949.

described in the previous section. As in the treatment with special relativity, one could use the general relativity "solution" and obtain consistent readings for the clocks in both systems, on the choice of one rocket as the base system; but, failing the existence of a preferred coordinate system in nature, it would be necessary to choose arbitrarily the base rocket, whose clocks would be ahead of those of the other rocket at the second rocket coincidence.

Our consideration of the proposed general relativity solution leads us to the conclusion that it does not give us a satisfactory resolution of the clock paradox. It is of interest that Professor Einstein apparently did not in his later years regard the general relativity solution to be of significance; in a conversation in 1952, he made the definite assertion to me that general relativity has nothing to do with the clock paradox.

5. CLOCKS AND ENERGY-FREQUENCY.

In addition to the clock paradox, there are other problems which arise when the Lorentz time transformation is applied to hypothetical clocks. One of these is the apparent inconsistency between the effect of relative motion on the energy-frequency ν and the slowing effect of motion on a clock.

Any particle or photon of energy E has associated with it a frequency ν, determined by $E=h\nu$, where h is Planck's constant. Suppose a particle of rest mass m_0 is measured, for its mass, by an observer moving with relative velocity v. Then, as we have seen in the previous chapter, section 6, this observer will find a mass $m=m_0/\sqrt{1-\beta^2}$; or, equivalently, the observer will find the energy and frequency to be given by $E=E_0/\sqrt{1-\beta^2}$ and $\nu=\nu_0/\sqrt{1-\beta^2}$ respectively, where E_0 and ν_0 are values for the rest observer. The equation $\nu=\nu_0/\sqrt{1-\beta^2}$ asserts that the effect of relative motion will be to *increase* the basic energy-frequency ν of a particle for the observer. But, contrariwise, the equation $\delta t_0=\delta t\sqrt{1-\beta^2}$ as applied to clocks, states that an observer of a clock will read an interval δt_0 on the clock moving with respect to him, compared with an interval δt on his rest clock. This means that the observer should find the moving clock to run slower, and hence have a smaller frequency, than his rest clock; for, if the moving clock has a slower rate, it must have fewer "ticks" when compared with a rest clock. The moving clock process therefore has a decreased frequency.

The contradiction between the *increase* in energy-frequency ν and the predicted *decrease* in clock frequency, as results of relative motion, is a striking one. The application of the Lorentz transformation to all processes, including clocks, implies that the time variable of all processes should be consistently affected by relative motion. How then can clock rates be decreased, and energy-frequency oscillations increased, as a result of relative motion? It is the case that a comparatively slower rate of a true "public" clock, as a result of relative motion, has never been observed, and we shall consider this fact in Chapter VIII in another kind of answer to the contradiction. Even, however, with full acceptance of the physical reality of the "slowing down" of a relatively moving clock, the conflict with energy-frequency increase may be shown to be removable. We here follow the treatment that was given by L. de Broglie, in the paper,[9] already cited in section 5, Chapter VI, in which he gave a complete discussion of his theory of wave aspects of matter.

De Broglie assumes that a moving body, as for example an electron, has a periodic oscillation of frequency $\nu_o = m_o c^2/h$ associated with it, where m_o is the rest mass of the body. An observer for whom the body is moving with velocity v should find a frequency $\nu_1 = \nu_o\sqrt{1-\beta^2}$ in accordance with the principle of slowing-down of moving clocks. On the other hand, since the energy of the moving mass is $m_o c^2/\sqrt{1-\beta^2}$, a frequency $\nu = (1/h)m_o c^2/\sqrt{1-\beta^2} = \nu_o/\sqrt{1-\beta^2}$ is to be expected. De Broglie remarks that this difficulty long puzzled him, until he saw a resolution in terms of a distinction between the periodic oscillatory phenomenon of the moving body and the phase waves (the "de Broglie waves" as we now say) of the body.

The body is moving with velocity v, for the fixed observer, while the phase waves have velocity c^2/v (as has been shown in our discussion of sec. 5, Chap. VI). De Broglie asserts the following theorem: The periodic phenomenon that is in motion and therefore has a frequency $\nu_1 = (m_o c^2/h)\sqrt{1-\beta^2}$ for the fixed observer appears to him to be constantly in phase with a wave of frequency $\nu = (m_o c^2/h)/\sqrt{1-\beta^2}$, itself propagated in the same direction as the moving body with a speed $V = c^2/v$. The proof of the theorem is as follows. Suppose that at time $t=0$ the phase of the periodic phenomenon of the moving body is in harmony with the phase of the ("de Broglie") wave of velocity c^2/v. At time t, the moving body has moved a distance $x = vt$ and the phase of the periodic phenomenon has been altered by

$$\nu_1 t = \frac{m_o c^2}{h}\sqrt{1-\beta^2}\,\frac{x}{v}.$$

The phase of that point of the de Broglie wave which is at the point of the moving body has been altered by $\nu t - \nu x v/c^2$; the second term enters because a displacement of x brings a variation of x/λ in phase, and $\lambda = (1/\nu)c^2/v$. Or, there has been a variation in the wave phase of

$$\nu\left(t - \frac{xv}{c^2}\right) = \frac{m_o c^2}{h}\,\frac{1}{\sqrt{1-\beta^2}}\left(t - \frac{xv}{c^2}\right).$$

[9] *Annales de Phys.*, ser. 10, *3*, 22 (1925).

Algebraic re-arrangement, using $t=x/v$, shows the right-hand side of the preceding equation to be equal to

$$\frac{m_0c^2}{h}\sqrt{1-\beta^2}\frac{x}{v},$$

which is the same as the alteration in the phase of the periodic phenomenon of the moving body.

The essential point in the foregoing resolution of the contradiction is that a slowing of the oscillations of the point phenomenon is consistent with an increase of the frequency of the extended waves, of velocity c^2/v. A distinction is made, then, between the oscillations of the mass itself, and the waves which are propagated through space, presumably as a consequence of the mass oscillations. However, without any kind of commitment to a model of the wave-particle that constitutes an elementary mass, we can state the distinction as being between the oscillations of an entity (the point particle) at rest in the moving coordinate system, and the *extended* wave oscillations of the particle; for it is the latter whose phase alterations when compounded of a time interval and a space interval are shown to be concomitant with the phase alterations of the contiguous point particle.

With a certain good-will, one can see the harmony of phase between the particle oscillations of decreased rate and the extended wave oscillations of increased rate as an elegant solution to our contradiction. The energy-frequency increase is derived for oscillations of a mass at a point in its coordinate system—that is, it is a purely time-like effect. But, presumably the mass particle is physically constituted as waves, and therefore it is the de Broglie phase waves which have the increased frequency; that is, it is the increased frequency of these waves which gives rise to the increase in energy E or mass m of a point particle, when the particle is observed at high relative velocity.

However, it may be asked: if mass particles are such that the decrease with motion of energy-frequency oscillatory rate is transmuted into an energy-frequency increase, then should not this same effect obtain for a clock? The clock, like the mass particle, is taken as located at a point in its coordinate system, and also the clock is made of matter; it is not unreasonable to think that the clock should therefore respond to relative motion as does a mass particle. Should it not also present a rate *increase* to an observer? The difficulty here is that we have no way of investigating the detailed response of a general clock to motion. We can expect that the de Broglie equation applies to the elements of the clock. But unless we consider a specific physical mechanism, we do not know whether there are any grounds for modifying the presumption of a rate decrease for clocks. Indeed, we hardly know what is implied by our assumption that the clock is "at a point."

The question at issue seems to be, whether or not we truly are justified in applying the Lorentz time transformation without reservation to *any* physical process, and we will discuss this question in later sections. With respect to the contradiction between energy-frequency increase and clock-rate decrease, we can say that de Broglie's treatment shows a way of removing the contradiction when we consider the well-confirmed relativistic increase of mass-energy; however, problems seem to remain when we consider the presumed behavior of clocks in general.

6. CLOCK RATES AND RELATIVISTIC PHYSICAL CHANGE.

In the presentation of the relativistic change of clock times we applied the Lorentz time transformation to a clock at rest in system S', moving with velocity v with respect to an observer's coordinate system S, and found the equation $\delta t = \delta t' / \sqrt{1-\beta^2}$. Or, using the sub-script zero in place of the prime mark, we can write,

$$\delta t = \delta t_0 / \sqrt{1-\beta^2}, \tag{1}$$

where the zero emphasizes the association of the variable t with the system S_0 in which the clock is at rest. The customary interpretation in special relativity theory is to consider δt_0 to be the time of a rest clock in S_0, as viewed by all observers, from any coordinate system, and to take δt as the time of a rest clock in the system S. The observer in S then finds $\delta t_0 = \delta t \sqrt{1-\beta^2}$, that is, that the moving clock runs slower than his clock.

The interpretation which gives us a slower rate for moving clocks may be emphasized by considering the difficulty which would face us if we essayed a different interpretation of Eq. (1). We might consider δt_0 to be the time reading that a clock at rest in S_0 gives for an observer also at rest in that system, and take δt to be the reading of that same clock for the observer who is at rest in S. Then we would interpret Eq. (1) to mean that the clock in S_0 gives a larger time reading, $\delta t_0 / \sqrt{1-\beta^2}$, for S observers than for S_0 observers. That is, that a moving clock runs faster than a rest clock.[10] We would however at once be in the difficulty that one and the same clock must present two different readings, δt and δt_0, at the same instant, to two different observers! There may, indeed, be an indefinitely large group of observ-

[10]At least one eminent student of relativity theory, E. T. Whittaker, does seem to have asserted that a relatively moving clock would run at a faster rate than one at rest. See his *History of the Theories of Aether and Electricity*, Nelson & Sons, Edinburgh, 1953, Vol. II, p. 42.

ers, each with a different velocity v with respect to the clock, and for each of these there must be a different reading. Specifically, the clock might be started at $t=0$, with the emission of an explosive light flash to signal its start, and run until δt_o, for its rest observer, with again a light signal from the clock to indicate that its mechanism is stopping. Observers in the various moving coordinate systems read the clock at the instant of the second flash, and are to find the various readings δt. Clearly, however, the clock can have only one reading at the second flash—and this reading is taken to be δt_o for all observers.

The public nature of a clock—the fact that it can be read by many observers at once—requires that it have the same reading, and in a sense the same physical behavior, for observers in all coordinate systems. The traditional special relativity theory doctrine does tell us that clocks run differently as a result of relative motion, but does not tell us that clock readings *of a given clock* vary from system to system. A given clock runs slow as a result of relative motion because it is compared with rest clocks which are running faster than the relatively moving clock. To repeat the matter in different words: each coordinate system may be thought of as containing clocks synchronized among themselves. When observers from different systems look at a given clock, there is no ambiguity or possible difference about the reading of *that* clock; but because the clocks in each system are running at different rates, in accordance with the Lorentz time transformation equation, relatively moving clocks seem to each observer to be running slower than his own rest clocks.

Now, however, it is also the case for the various observed relativistic physical processes that each observer does find events to be different, depending on his velocity relative to the events. Thus, a mass m_o, at rest in the system S_o, is observed as a mass $m=m_o/\sqrt{1-\beta^2}$ by an observer in the system S moving with relative velocity v. Likewise, photons of frequency ν_o from a source at rest in S_o are found to have various frequencies ν by observers in systems S moving with various velocities relative to S_o. And, similarly, mesons at rest in S_o present different life-times to various observers S, and elementary particles at rest in S_o are found to have different de Broglie wave-lengths by observers in systems S that have differing velocities relative to S_o. In all of these cases, in contrast to the situation with respect to clocks, it is as if Eq. (1) were interpreted so as to give the time variable reading δt, for the phenomenon in question, to the observer S, and the variable δt_o for the phenomenon only for the rest observer in S_o. And yet in all of these observed physical confirmations of the Lorentz time transformation there is no possibility of paradox or contradiction, because any *one* instance of the phenomenon in question is being observed from some *one* relatively moving coordinate system. In each observation, a definitely given physical interaction, with one observer's apparatus, is involved. A given electron at rest in S_0 can interact only with a given mass-measuring electrical field, at some relative velocity v. Any *one* photon from the source at S_0 can interact only with an observer who has some given velocity v with respect to S_0, even though other photons from the same source may be interacting with observers that have other velocities relative to S_0. Likewise, the de Broglie wave-length is determined by the diffraction of a group of particles in a given apparatus; or, the decaying meson interacts with a given group of atoms or electronic counters. In all of these instances, in contrast with clocks, there is no "public" character of the event observed, such that it may be simultaneously observed from any number of different coordinate systems; rather, there is a single relative velocity that determines the nature of the transformation for each phenomenon.

The observed relativistic changes of physical quantities, like mass, frequency, and wave-length, taken with the presumed *relative* slowing of clock rates, give an insight into the central role of clock-rate changes in the usual interpretation of relativity theory. As we have pointed out, a clock at rest in a given coordinate system is taken to have a single, invariant *reading* for all observers, but because clocks in different coordinate systems have different readings, each observer finds clocks in systems other than his own to run slower than his own. Analogously, we may presume that other physical events than clocks are independent of the observer's coordinate system: that they have a tempo and structure that is inherent to them and their coordinate system, but that they are found to vary in physical characteristics when they are observed from relatively moving coordinate systems, because of the variations in time (and clock) rates among the various observers' systems. Thus, the electromagnetic waves from a given source may, on this view, be taken as objectively existent in a way that is independent of observers, but because the clocks of system S run differently than do those of S_0 (or, more generally, because time rates are different in S than in S_0) the waves appear differently to S observers than to S_0 observers. The same point of view may be taken toward the other observed relativistic changes that we have discussed: the mass or the wave-length of a relatively moving particle, or the decay time of a moving meson, take on different values as observed from different coordinate systems, as a result of the time transformation between systems.

Metaphorically speaking, we might express this view of relativistic physical changes by saying that a given set of physical events is in itself independent of observers' coordinate systems, but that the time-rate peculiar to the clocks and measuring apparatus of each coordinate system form a pair of distinctive eye-glasses. The hue of each pair of glasses varies with the velocity relative to the observed system, and each relatively moving observer's spectacles present him with different values of mass, frequency, etc. for a given set of physical events.

For brevity, we shall refer to the interpretation that we have been discussing for relativistic changes as the "clock-rate change" view, for its distinctive feature is that relative changes in clocks and time rates generally—rather than in specific physical phenomena—are the essential elements in the relativistic changes of the physical world.

Even the contraction of moving lengths may be regarded as a consequence of the transformation in time scales between relatively moving systems. The contraction follows from the Lorentz equation (see Eq. (3), sec. 1, Chap. VI),

$$\delta x' = \frac{\delta x - v\delta t}{\sqrt{1-\beta^2}}. \tag{2}$$

A length $\delta x'$ is taken to be at rest in a system S', moving relative to system S with velocity $v = v_x$. A measurement of the length by an observer in S must be made with $\delta t = 0$; that is, both ends of $\delta x'$ must be measured simultaneously, according to S clocks. With $\delta t = 0$, Eq. (2) gives for the S observer, in terms of his variables, $\delta x = \delta x' \sqrt{1-\beta^2}$; or, the S observer will find the length $\delta x'$ to be shortened by the factor $\sqrt{1-\beta^2}$. Conversely, however, application of the transformation equation,

$$\delta x = \frac{\delta x' + v\delta t'}{\sqrt{1-\beta^2}},$$

will give the result that an observer in S' who makes a measurement with $\delta t' = 0$ will find a length δx at rest in S to have the contracted length, $\delta x' = \delta x \sqrt{1-\beta^2}$. The S and S' observers do therefore find different values for one and the same rod, but the values are as each determines them by his own measuring stick, and there is no contradictory situation of one measuring rod having simultaneously to present two different readings; for again, unlike the "public" clock reading, each "length of rod" is referred to the measuring apparatus of a specific coordinate system. The fact that different observers find different length values for the same rod can be taken as a consequence of the relativistic difference in simultaneity of events in different coordinate systems. Measurement of a length in S', made with $\delta t' = 0$ by the rest clocks of the S' observer, is made at two different times, by the reckoning of an S observer, since his clocks give a different indication of simultaneous ($\delta t = 0$) measurements.

The "clock-rate change" interpretation of the physical changes that are predicted by the special theory of relativity seems to be the one that is most widely accepted by relativity theorists, and the interpretation may be generally taken as confirmed by the many impressive successes in physics of special relativity theory. Still, we have seen that this view carries some embarrassments with it—notably the problem of the clock paradox, where the solution by way of appeal to a preferred coordinate system is (at least in the minds of many) a little less than completely convincing. Certain other difficulties will be discussed in the next chapter, and also, in that chapter an alternative interpretation of relativistic changes will be suggested. But first we shall again discuss the relativistic changes in meson decay rate, in the final section of this chapter.

7. CLOCKS AND MESONS.

The observed increase of mu-meson life-times with relative motion, described in section 4 of the preceding chapter, gives a direct illustration of the effect of motion on the life-time of a physical process, and an extrapolation is often made from the meson effects to the slowing of a relatively moving clock. We shall now see that this extrapolation can very nicely further elucidate the "clock-rate change" point of view.

A meson is of course not a "public" clock, for the meson decays with respect to an atom, or group of atoms, or an electronic counter. The mu-meson in its decay emits a positron or electron, and this particle might have an initial interaction with, say, an electronic counter; the meson would have some velocity with respect to this counter, and its relativistic life-time increase would be determined, on the average for a group of mesons, by that velocity. In practice we say that the meson's life-time is determined by its velocity relative to the particles with which its decay particles interact. Clearly, a measurement of life-time for a given meson could not be made from a number of systems, since the initial physical interaction of the meson's decay particle can be with respect to a counter or atom in only one system. Hence, the meson cannot be observed as a clock from many systems in various states of relative motion; and hence, we can write $\tau=\tau_0/\sqrt{1-\beta^2}$ for the observed meson life-time τ, with no paradoxical problem of there being different τ values for observers in coordinate systems with different v values with respect to the meson.

A hypothetical moving clock can be read by different observers. We may relate the clock and meson times by assuming that in the rest system S_0 of the meson there is also a rest clock. A laboratory observer in system S finds a life-time τ for the meson, which has a velocity v with respect to him. For an observer in S_0, at rest with respect to the meson, the meson should however have a rest life-time τ_0; and in accordance with the equation $\delta t_0=\sqrt{1-\beta^2}\,\delta t$ that observer should find that his S_0 clock reads τ_0 rather than τ when the meson decays. There is, then, complete agreement between the observed meson effect and the slowing of clocks: all physical processes are decreased in time rate as a result of motion, and therefore both the clock and meson "life processes" in S_0 are slow, as measured by a laboratory clock in S. The smaller clock reading, δt_0, results from a rate decrease, just as a longer meson life-time, τ, as measured by an S clock, results from slower "life processes" in the moving meson.

Or, we can say that the processes of the meson and of the clock, both at rest in S_0, are in themselves unaffected by relative motion of an observer's coordinate system. But, because each relatively moving coordinate system has a characteristic time rate, the S_0 meson and clock appear to have different rates when they are judged by the clocks of various observers' systems. Specifically, the S observer finds, by his clocks, that the S_0 clock runs slow and that the meson at rest in S_0 requires a longer time for decay than would a meson at rest in S. And, we may argue, the meson effect is particularly convincing evidence in favor of the "clock-rate change" view, because the meson decay is not a periodic physical process but one that occurs only once for a given meson, and is actually of varying length from one rest meson to the next. A characteristic tempo for *all* of the processes in a given coordinate system is therefore suggested; and the tempo of any one coordinate system apparently is related to that of another system by the Lorentz time transformation equations. The assumption of change in clock rate with relative motion seems, then, to be all but directly confirmed.

A peculiar difficulty arises, however, on further consideration of the role in meson decay of the relative velocity between a meson and the matter with which its decay particles interact. We have been considering mesons in system S_0 to move at relative velocity v with respect to, say, electronic counters at rest in S. These counters record the decay of the S_0 mesons, and an average meson life-time τ is observed. Let us now suppose that in the moving system S_0 there is not only a hypothetical rest clock but a hypothetical observer, at rest in the system. Further, we assert, i) that the relative velocity v between the mesons and counters determines the life-time change, in accordance with the equation $\tau=\tau_0/\sqrt{1-\beta^2}$, and, ii) that by the principle of relativity it is immaterial whether the counters or the mesons be considered to be in motion. But then we have the following result.

The meson does not decay with respect to rest counters, for the S_0 observer. Rather, his description of events must be that the rest meson's decay was with respect to an electronic counter which rushed by at a high velocity v. Also, it does not matter whether the counter is described as moving with respect to the meson, or the meson with respect to the counter. Hence, the S_0 observer *should calculate a relativistic life-time for the meson,* and should expect to find a decay time τ, not a time τ_0. That is, to be specific, if the decay of the meson is signaled to the S_0 observer by a light flashing in the S counter when that counter receives the decay electron of the meson, the time of the flash should indicate a life-time τ for the meson to the S_0 observer. However, we have already agreed that in this situation the S_0 clock records a mean life-time τ_0 for the S_0 mesons.

The customary "clock-rate change" point of view forbids our saying, then, that it is the relative velocity of physical interaction between meson and surroundings which determines the meson life-time. If it were this velocity, we would expect that both observers in S and S_0 would find a life-time τ for the mesons that interact with the moving counters, because both observers would consider mesons to decay with respect to matter with which there is a relative velocity v. And since both S and S_0 clocks would give the reading τ, it would then be necessary to regard the rates of the rest clocks in S and S_0 as unaffected by the relative motion between the two coordinate systems.

The preceding discussion emphasizes that on the "clock-rate change" view, relativistic physical change does not in any way rest on the existence of physical interaction at relativistic speed. Indeed, we saw that assumption of physical interaction as a criterion for there being relativistic effect led to a contradiction in the system of clock-rate changes. *Relative velocity alone* determines relativistic effect on this view: we must, in the situation we have been discussing, consider the mesons (and clocks) of system S_0 inevitably to be slower in rate than those of system S, regardless of any physical interaction between the two systems. Our considerations have been purely hypothetical, it is true, in that we know of no actual physical situations in which observers move along with mesons, at speeds where relativistic effects are significant; but also, it must be remembered that all considerations of relativistic effects in actual "public" clocks are entirely hypothetical.

An alternative view is that relativistic changes in a physical system are to be associated with physical interactions of the system. Then, for example, clocks in *both* systems S and S_0 must read τ for the average meson decay time if the meson in its decay interacts with a counter moving with speed v, and we dismiss the existence of a transformation of clock times between the two systems, on the grounds that the internal operations of clocks do not involve any physical interactions at relativistic velocities. The feasibility of adopting an interpretation of relativistic time change which includes this view of clock and meson-decay rate changes will be examined in detail in Chapter VIII.

Chapter VIII
Lorentz and Clausius Processes

In this third chapter devoted to time and relativity theory we consider a major difference between time transformations of clocks and of observed relativistic processes. A somewhat novel approach *that is altogether speculative* will be presented, and this approach must not be considered as representing an established trend of thinking in current physics. In this respect, the present chapter (and some parts of the clock problem discussions of the preceding one) differs from the other chapters of the book; for in general the results and concepts of physics which we have discussed are among those that are commonly accepted by contemporary physicists.

I shall suggest that change of clock rates with motion is not essential to special relativity theory, and further, that a distinction should be made between two kinds of natural processes with respect to the applicability of the Lorentz transformations. With this distinction, the "clock problems" of the special theory of relativity are readily solved, but in a manner quite different from that in the currently accepted clock-rate change doctrine. The distinction is on the level of interpretation of the Lorentz time transformations, and it does not involve any change in the basic equations of special relativity theory or in the way the theory is used to-day by physicists in studying natural physical processes. But the distinction does entail a radical change in the application of the theory to hypothetical clocks. Specifically, it is suggested that the origins of relativistic time changes are in *physical interaction* at relativistic velocities rather than in relative motion itself.

1. THE RESTRICTION TO PHYSICAL INTERACTIONS.

In discussing the relativistic time effects of Chapter VI we emphasized that in all cases the process undergoing a time transformation was in a physical interaction with the observer, or with his measuring apparatus. The interaction may be through a physical field, as when an electron mass is determined by observing the electron's deflection in an electric field with respect to which it has a velocity v. Doppler frequency changes for photons involve a direct physical interaction between the photon and the atoms in the eye or recording apparatus of the observer moving with velocity v with respect to the light source. The de Broglie wave-length of a group of particles is measured with respect to the crystal or other diffraction apparatus which presents alternative path lengths of different magnitude to various parts of the de Broglie waves, there being the same kind of direct interaction between the particles and the observing mechanism as in the measurement of photon wave-lengths. We have already emphasized that in studies of relativistic changes in meson life-times the decay particles of a given meson interact with the atoms of the surrounding medium, with respect to which the meson has the velocity that determines its life-time.

In all observed instances of relativistic change in temporal characteristics there is a definite "physical situation" involving both the process which is undergoing the change and the observer of the process; more specifically, the process is only observed by virtue of its interaction at high velocity, as a process, with physical elements of the observer's coordinate system. In contrast, the hypothetical clock of relativity theory is not involved in a definite physical interaction with the observer of a presumed relativistic change in clock rate. The act of reading the clock does not involve any element of the clock mechanism in a high velocity interaction, even though the clock is moving at such a velocity with respect to the observer. For this reason there is nothing unique about the time transformation on a clock rate in a given physical situation, and the clock is "public," and can at any instant be read by any number of observers.

Suppose a clock to consist of spring, cog wheels, escapement, dial, hands, and various connecting parts. The clock, moving with a velocity close to that of light, is read by a stationary observer. We clearly would expect that a Lorentz transformation must apply to the entities which form an interaction between the clock and observer, and which enable the latter to read the clock. Thus, if the clock dial were luminescent we would expect a Doppler shift to obtain for the photons received by the observer from the dial. But the internal clock parts could hardly in any way be undergoing relativistic interactions as a result of the high velocity between clock and observer. The motion between the clock and its observer is not a relativistic motion that involves the essential clock process, namely, the spring and cog-wheel motions. Or, put another way, the clock process is not in relativistic interaction with the observer; the *reading* of the clock is a relativistic physical interaction, for which we expect the appropriate relativistic transformations, but the process which determines the clock rate is physically independent of that reading.

We now suggest that the application of the Lorentz transformation equations be made only to processes which are in physical interaction with an observer (or his apparatus). In general, the transformations will be of interest only when the interaction involves a high velocity between the process and the observer. We might call these processes, to which the transformation equations are applicable, *Lorentz processes.* A Lorentz process, then, is one such that observation of it involves the process itself in a physical interaction with the observer, generally at high relative velocity. The space and time variables of a Lorentz process are transformed, and associated with each other, in the manner that is prescribed by the Lorentz transformation equations.

A space or time variable of a Lorentz process may take on any value that is determined by the Lorentz transformations, because the process, in any physical situation, is in a definite physical interaction with an observer or physical equivalent. As we have seen, mu-mesons for example may have varying life-times τ, depending on their velocity with respect to the physical medium with which their decay products interact. Physical properties of Lorentz processes, dependent on space or time variables, likewise take different values for different observers, as in the variation of mass-energy with velocity. In a word, the *physical characteristics* of Lorentz processes are genuinely and physically changed as a result of the interaction and relative motion between the process and its observers.

A second kind of process does *not* involve essential interaction with an observer when a "reading" is made of the process. We shall refer to these as *Clausius processes.* An observation of a Clausius process does not involve the process in an interaction such that the process is excluded from being observed, in the same physical manner, by observers in other coordinate systems also. Hence, the Clausius process may be simultaneously observed, by similar physical means, from any number of different coordinate systems. And further, the Clausius process therefore cannot undergo physical relativistic transformations for different observers, because if it did it paradoxically would have to present different characteristics at the same instant. The public clock is such a process, and we have seen even on the usual "clock-rate change" point of view, how it must be taken as presenting a single reading, or time rate, to observers in all coordinate systems, with relativistic transformations obtained through the differences of clock rates in different coordinate systems.

Because the Clausius process is not, in itself, in physical interaction with its observer, and because the process therefore cannot take on a definite physical transformation corresponding to a given state of physical interaction, I suggest that the relativistic Lorentz transformations are not applicable to these processes. In general, Clausius processes are of complex systems, and the internal interactions of the system are not physical interactions which are involved in the *observation* of the Clausius process. And also, the Clausius process is internal to a system, with the result that relativistic velocities are not involved, whatever the velocity of the total system may be. If the Clausius process is in any way a temporal process, its time rate will not, we suggest, be affected by relative motion between itself and an observer. Rather, the Clausius process is *invariant* under a transformation between coordinate systems moving with relative uniform motion. Complex systems like biological organisms, as well as public clocks, would be Clausius processes, and their aging rates would not be affected by relative motion with respect to an observer.

As an example of a Clausius process whose mechanism is somewhat simpler physically than that of a conventional clock, we consider a clock made of two chambers with a small connecting tube through which gas can leak from one chamber to the other. One chamber, *A*, is originally filled with a gas at a pressure P_i, and the other chamber, *B*, is originally at zero pressure. When a stopcock in the connecting tube is opened, gas slowly passes from *A* into *B*, and a gauge on *B* registers the pressure P_B. The rise in P_B is calibrated by a standard clock, so that the increase of P_B from zero to a final pressure P_f will serve as a clock. The process of gas transfer from *A* to *B* is a Clausius process. An observer in fast relative motion with respect to the gas clock would not in any way be in physical interaction with the gas molecules, and a reading of the pressure gauge, by visible light, or radio waves, would not involve any interaction of the clock processes as such with the observer. Our suggestion for Clausius processes requires that the time rate of change of pressure P_B is in fact independent of the motion of an observer with respect to the chambers *A* and *B*, and that two such gas clocks, if they have the same time rate when compared at rest, will also have the same rate if they are compared when in uniform motion relative to each other.

Clausius processes can in many instances be characterized as being thermodynamic processes, to which thermodynamic variables, and specifically an entropy function, are applicable. We shall discuss in a later section the role of entropy in the relativistic invariance of Clausius processes; we note here, however, that the name of Clausius seems appropriately descriptive because of his formulation of the entropy concept.

We do not wish to suggest that the relativity theory equations are not valid for the internal physical interactions of a Clausius process. The Lorentz transformations can be expected to be applicable to any physical process, and we saw in discussing mass and velocity, in section 6 of Chapter VI, that the ordinary inertial resistance of mass at even a vanishingly small velocity may be considered to be a consequence of the Lorentz time transformation. But in general the Lorentz equations give no significant difference from results based on non-relativistic equations unless the relative velocity of interaction v is such that v^2/c^2 is a non-negligible quantity. The internal interactions of a Clausius process would not in general involve any high relative velocities, and hence there would be no "relativistic effects," even though the system or process as a whole might be moving at high velocity relative to an observer. One can conceive of a situation in which *internal* relativistic effects do need to be taken into account in a Clausius process. The molecular velocities in the gas clock, for example, are usually such that relativity effects need not be applied in the molecular collisions. But the gas might be so hot (of the order of 10^8 °K or greater) that it would be necessary to take into account relativistic effects between the molecules, since their masses and relative velocities on collision would then be as given by relativity theory equations.

The suggestion that relativistic transformations do not apply to Clausius processes is in strong contrast to the standard "clock-rate change" doctrine. The latter requires, for example, that the molecules in a gas clock, as described, would move more slowly from chamber A to B if the clock were in system S_0, moving with respect to S, than if the clock were in S. The fact, however, that actually there would be no physical interaction between the clock processes in S_0 and the observer in S gives some immediate physical appeal to the suggestion of time invariance for the Clausius processes. Also, it is perhaps clear that the various clock problems which we discussed in Chapter VII will be readily solved on the basis that public clocks are Clausius processes whose time rates do not change with motion. We shall later discuss these clock problems from this point of view. First, however, we must consider the question of the relation of the constant velocity of light to the proposal of time-invariant Clausius processes.

2. THE VELOCITY OF LIGHT.

In our derivation of the Lorentz transformation equations (sec. 1, Chap. VI) we required, as an initial assumption, that relatively moving observers find the same speed c for light coming from a given source; this assumption, taken with the lack of any evidence for a physically defined preferred coordinate system by which states of absolute rest or motion may be determined, forms the keystone of special relativity theory. We say, in the derivation, that a light signal spreading from a spherical source (see Fig. 5, p. 83) must appear both to an observer S and an observer S' to spread out as a uniform spherical wave, in spite of the fact that S and S' have a high relative velocity with respect to each other. This requirement, which is tantamount to the existence of the same light speed c for all observers, was fulfilled by changing the length and time scales—the rods and clocks—of S' with respect to those of S; that is, was fulfilled by requiring the changes on length and time variables that are the Lorentz transformations.

It is generally taken in derivations of the Lorentz equations that a travelling light front is a unique physical entity, having a definite and single physical location. With this assumption, there must be a "giving" on the part of the space and time variables which measure the progress of the wave-front, if relatively moving observers S and S' are to find the same light speed c for the front. The "clock-rate change" interpretation of the Lorentz equations requires just those changes in clock rates from one system to another, and hence also in measured lengths, which will ensure that observers S and S' will *each* find the single wave front to be travelling outward as a sphere with velocity c. We employed the unique light front signal in the customary manner, in our derivation in Chapter VI. Now, however, I wish to propose an alternative assumption. Instead of taking the light front to be one and the same for all observers, we will apply the Lorentz transformation equation to the light front in a way such that the space and time variables of the photons of the light signal are transformed, rather than the observers' measuring rods and clocks. We shall say that the derived transformations which guarantee that both S and S' observers find a speed c for light from a given source are transformations on the variables of the photons which interact with the observers. We shall then be using our "physical interaction" interpretation of the Lorentz transformation equations. The physical characteristics of the photons from a source are changed, with change in the velocity of the observer with whom the photons interact. Specifically, the wave-lengths and frequencies of the photons vary, in accordance with the Lorentz equations, so as to maintain a constant value for the velocity c, the product of wave-length and frequency. The constancy of c is a result, then, not of difference between the time rates of all the processes of one system compared with another, but of the alterations in the physical properties of interacting photons.

This change in point of view is illustrated in Fig. 10. The customary, unique-light-front interpretation is shown in the sketch of Fig. 10(A). The coordinate origins O and O' of systems S and S' are together at the time $t=t'=0$, when a light signal is emitted from the spherical source A. The source is fixed at O and is therefore stationary in S. At a time $t=x_1/c$, observers in S receive signals at x_1 and $-x_1$.

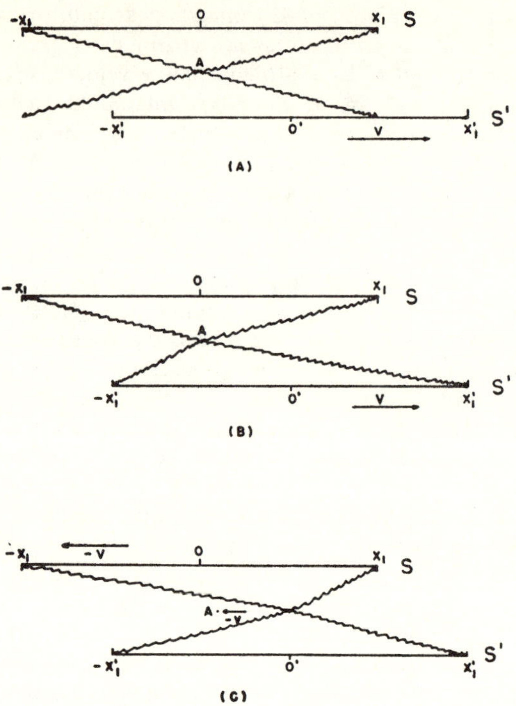

Fig. 10
Light from source A, at rest in S, is emitted when O and O' are in coincidence, at time $t=t'=0$.
A) The light photons are assumed to have identical propagation properties for S and S' observers.
B) System S is taken to be at rest. Photons are assumed to move in the manner that gives a speed c with respect to the system with which they interact; therefore the photons from source A which interact with S' observers have propagation characteristics different from those of photons from A which interact with S observers.
C) As in (B), except that S' is taken to be at rest.

At the same moment (for S) that it strikes x_1 and $-x_1$, this same wave front is beyond the point $-x_1'$ of S' and has not yet reached $+x_1'$, because S' is in motion during the time t in the $+x$ direction. For S', however, the wave front when at x_1 is at a point $x'=(x_1-vt)/\sqrt{1-\beta^2}$ at a time when the S' clock reads $t'=\dfrac{t-x_1v/c^2}{\sqrt{1-\beta^2}}$, and the wave front at $-x_1'$ is at a point $x' = (-x_1 - vt)/\sqrt{1-\beta^2}$ at an S' clock time $t' = \dfrac{t + x_1 v/c^2}{\sqrt{1-\beta^2}}$; the change in the space and time measurements of S' from those of S ensures that the identically same wave front will be found by both S and S' to have a propagation speed c in all directions.

Our "interaction point of view" with respect to light speed is illustrated in 10(B). We consider the same situation of systems S and S' as before, with light emitted from the source A at $t = t' = 0$ when O and O' are at a common point. But now we consider the time of the clocks in S and S' to be unaffected by relative motion, taking the clocks to be Clausius processes. For S, as before, light signals come a distance x_1 from the stationary source, in time $t = x_1/c$. For S', the source is in motion; this motion however should not affect the speed of propagation of light to an S' observer, and, since A was at O' when the signal was emitted, the light should reach x_1' at $t' = x_1'/c$. Since $x_1' = x_1$, $t' = t$. S and S' clocks are running at the same rate, so, the photons from A which strike x_1' must travel a longer distance, from the S point of view, than do the photons which strike x_1. We no longer take the spreading wave-front from the source to be physically unique. Instead, we say that different parts of the wave-front travel differently; specifically, any given photon moves so that it will reach the coordinate system, *in which it is observed,* with a velocity of magnitude c. The Lorentz transformations *are on the photons with which observers interact,* not on the clocks and rods with which the observers make measurements.

Figure 10(B) shows us that a photon which strikes $-x_1'$ covers a smaller distance, to the S observer, than do photons which strike $-x_1$. However, from the S' point of view, Fig. 10(C), taking S to be the moving system, it is the photons which strike x_1 and $-x_1$ which must cover shorter and longer distances, respectively, than do the photons which strike x_1' and $-x_1'$. The S' observer considers the source and the system S to be moving to his left at a velocity $-v$; but since the source was at O' when the light signal was emitted, for him photons travel equal distances in coming to x_1' and $-x_1'$. In proper relativity fashion, either observer may consider the other to be in motion, and either may consider that it is the other for whom the photons must move differently as they go to right or left. There is no possibility of paradox because *any one given photon* of interest interacts with S or S', but *not with both.*

The transformations, on the space and time variables of the photons, which will require that they move with a velocity c with respect to either the S or S' observer are just the Lorentz transformations, which may be derived and applied to photons exactly as has been done in

Chapter VI. And, because photons are "Lorentz transformed" as required by the Doppler equation, they of necessity move with a velocity c with respect to the observer. Further, the Doppler frequency shift for electromagnetic waves is completely confirmed observationally (see sec. 3, Chap. VI), and is therefore a Lorentz process which is directly observed. This means that by ensuring the constancy of light speed for all observers by means of Lorentz transformations on photons, we apply the transformation equations in the same manner as with other observed relativistic processes, and we get immediate confirmation of our application in the observed Doppler shifts. In contrast, in order to ensure constant light velocity by systematic changes in clocks (and rods) as a result of relative motion we must require clock changes as a result of relative motion without essential physical interaction. There is no empirical evidence for these changes, and, as we have seen, there are difficulties that result from admitting these clock changes.

On our suggestion of electromagnetic waves as Lorentz processes, and clocks as time-invariant Clausius processes, no observer can measure a vacuum velocity other than c for any photons from a given source. It is the case, however, that conceptually velocities of light greater or less than c will appear for observers in systems other than the one in which measurement is being made. Thus, referring again to Fig. 10(B), the observers in S and S' each consider that light requires a time $t=t'$ in reaching x or x'; to S' the light left its source at $t'=0$, when the source was at a distance x_1', which is the same as the x_1 distance in S. The S observer would regard S' to have moved a distance $vt=vx_1/c$ during the time that light travels from A to x_1, and that hence, when a photon from A reaches x_1', it will have had to travel a distance $x_1'+vt=x_1+vx_1/c$. Since it travels this distance in time $t=t'$, its apparent velocity to S will be $(x_1/t)(1+v/c)=c(1+v/c)$. Similarly, the photons from A that strike $-x_1'$ in S' at t would seem, to S, to move with a velocity $c(1-v/c)$. The S' observer (see Fig. 10(C)) would make similar statements about inferred photon velocities in S. This conceptual existence of vacuum light speeds other than c is analogous to what S and S' observers conceptually expect, on the traditional unique light front view, for each other. Thus, the light sphere that is spreading uniformly from A for the S observer should look skewed to the S' observer, if he does not consider the Lorentz transformations that are applied to change the clocks and rods of S' with respect to those of S. These transformations applied directly to the photons describe the fact that the photons which interact with the S' observer literally do move differently (with different frequencies and wave-lengths) from photons from A that interact with an S observer, with the result that S' does measure a light speed c.

The key element in our reconciliation of clocks' being unaffected by relative motion with the hypothesis of the same measured light speed for any observer is the varying behavior of photons from a given source with respect to observers moving relatively to each other. We require that photons, as it were, must "know" when to hurry up so as to interact at velocity c with an observer who is receding from the source, and also must know when to slow down so that their velocity of interaction will also be c if the observer is moving toward the source. This adjustment becomes less mysterious if we state it as being one such that the medium—i.e., space—will only permit interaction of photons with atoms when the relative photon-atom velocity is c.

We know that in any event the photon is not a definitely localized physical entity with invariant properties. The Doppler shift shows us the variation that is possible in the properties of a photon. Interference experiments indicate that the photon is spread out along its direction of propagation to an indefinite extent: with respect to the question, "What is the length of a photon?," observations give no definite answer.[1] Our present knowledge of electromagnetic waves or photons does not, therefore, in any way forbid that light propagation be physically transformed so as always to be at a speed c with respect to an observer.

The problem of light-speed invariance clearly needs further detailed study. With our physical-interaction requirement for relativistic transformations, the invariance of c is a question that centers on the properties of space as a medium for electromagnetic waves and on the nature of the interaction of these waves with matter. With the traditional viewpoint of clock-rod change with motion, the invariance of c does not appear to be so readily available for further study and understanding, since it rests on hypothetical changes of clock rate that occur with relative motion, independent of any physical interactions.

3. THE CLOCK PROBLEMS.

If we accept the suggestion that public clocks are Clausius processes, as defined, and therefore unaffected in rate by relative uniform motion, we at once solve the problem of the clock paradox. We then would not expect, for example, that the clocks in the rocket situation that was presented in the preceding chapter would differ at all in their readings on the second coincidence of the rockets at the point P. Likewise, the brother Paul, in Langevin's "twin paradox," would find

[1] See discussion in E. T. Whittaker, *History of the Theories of Aether and Electricity*, vol. II, Nelson, Edinburgh, 1953, pp. 94-95.

himself and Peter to have aged equally when he returns from his high speed journey in space. In general, the rates of change of clocks and biological organisms would be independent of uniform motion relative to an observer.

Physical acceleration might well have effects on the rates of clocks and other complex systems, but these effects would be problems for detailed analysis, and they are not relevant to the clock problems of special relativity theory. Also, our definition of Clausius processes does not in itself exclude the possibility of effects of gravitational fields on clock rates, as required by general relativity theory. The independence of clock rate from uniform motion does raise a question of whether time should be characterized as being "absolute" or "relative"; the question will be discussed in section 5 of this chapter.

The increase in meson life-times becomes consistent with clock behavior, after we adopt the distinction of Lorentz and Clausius processes. A clock attached to a laboratory counter with which the decay products of rapidly moving mesons interact, and a hypothetical clock in the meson's coordinate system, will *both* record, on the average, the same life-time $\tau=\tau_0/\sqrt{1-\beta^2}$, since the clock rates are not affected by uniform motion; this result we might expect, since in the coordinate system of either the counter or meson there is a primary interaction *at high relative velocity* between the meson, through its decay products, and the counter. Similarly, a rest decay process, in which meson-decay products interact with atoms or counters with respect to which the meson is relatively at rest, will be recorded as indicating a life-time τ_0.

Our consideration of clocks as time-invariant, Clausius processes readily solves the clock problems because those problems arise only with the introduction in relativity theory of rate changes in hypothetical clocks. There are no "clock problems" in relativity theory as such; that is, in the Lorentz time transformation equation and the *application of that equation to the time variables of processes in physical interaction at relativistic speeds.* Rather, as is shown by the examples of relativistic time transformations which we have discussed, that application beautifully confirms that rates of a process are different—there is a difference in the values of t and t'—depending on whether observers in relatively moving system S or S' interact with the process. On our "interaction view," the hypothetical non-interacting clock may be entirely dropped from consideration in the discussion of relativistic time changes.

4. COMPLEX SYSTEMS.

The traditional point of view with respect to clocks and special relativity theory has held that the time rate of any process will be found to be decreased by a factor $\sqrt{1-\beta^2}$ when that process is observed from a coordinate system moving with a relative velocity v. Our suggestion of the distinction between Lorentz and Clausius processes in no way questions the validity of the Lorentz transformations when applied to a process that is in physical interaction with the relatively moving observer. For processes which do not involve internal relativistic velocities and which are independent of observation from the relatively moving coordinate system, we suggest that application of the Lorentz transformation is unwarranted, and that it is a source of difficulty if it is made.

A further way of studying the validity of application of the Lorentz transformations is by investigation of the detailed mechanisms of mechanical clocks. One might, for example, consider some simple physical clock mechanisms, as rolling balls, falling bodies, or oscillating masses, and apply the transformation equations to the various internal clock motions.[2] One should, on the clock-rate change view, come to the final result that a clock moving at relative speed v runs slower by the factor $\sqrt{1-\beta^2}$. In some cases one does gain this result, and in other cases not. It is, however, difficult in most instances to judge how significant the results are, because, even in the framework of the traditional point of view in relativity theory, we do not always completely know what the effects of motion will be on materials or mechanisms.

We can take another approach to the problem of the effect of relative motion on the internal processes of a clock if we use not a simple mechanical mechanism but rather a thermodynamic one as the clock process.

Some attention has been given[3] by relativity theorists to the question of the effect of a Lorentz transformation on those properties of highly complex systems, as say a gas containing billions of molecules, which are considered by the methods of thermodynamics and statistical mechanics. The results of these studies lead to a strong argument against any effect of uniform motion on the time rate of a clock (again, when there is no essential interaction between the clock process and moving observer). In section 1 of this chapter we discussed a gas clock as an example of a Clausius process. The molecular motions and interactions, as a result of which gas escapes from one chamber to the other, were suggested as being independent of relative, uniform motion of the clock system. We shall now present the argument for invariance of

[2]Herbert Dingle has discussed the difficulties that arise with the application of the Lorentz transformations to the mechanisms of a literal clock, and has argued against the physical validity of the Lorentz time transformation for all clocks. *Nature, 144,* 888 (1939); *146,* 391 (1940). *Amer. Journ. Physics, 10,* 203 (1942); *11,* 228 (1943).

[3]See M. Planck, *Ann. der Physik, 26,* 1 (1908); W. Pauli, *Encyckl. der Math. Wiss.,* Band 2, Heft 4, Leipzig, 1921; R. C. Tolman, *Relativity, Thermodynamics, and Cosmology,* Oxford Univ. Press, 1934; C. Eckhart, *Physical Rev., 58,* 919 (1940); M. von Laue, *Die Relativitätstheorie,* Band I, 5 Aufl., Braunschweig, 1952, pp. 145-179; B. Leaf, *Physical Rev., 84,* 345 (1951); P. G. Bergmann, *Physical Rev., 84,* 1026 (1951); G. A. Kluitenberg and S. R. deGroot, *Physica, 20,* 199 (1954); *21,* 148 (1955).

the time rate of such a clock or of any non-interacting thermodynamic system, in terms of the results that have been accepted in special relativity theory itself for thermodynamic systems.

In Chapter II we discussed an important thermodynamic quantity of a system, the entropy. Let us denote by the symbol S the entropy of a system G, e.g., a quantity of gas in a container. We shall consider G to be at rest in the coordinate system Σ. We then ask the question, What is the entropy S' of G for an observer who is in a coordinate system Σ', moving with uniform velocity v with respect to Σ? The generally accepted answer is that there should be no change of entropy as a result of a Lorentz transformation; that is, $S'=S$.

The preceding conclusion has been reached in various ways by different workers in special relativity theory, and, in detail, it is necessary that other quantities, as temperature and heat, in terms of which entropy may be defined, have Lorentz transformation properties which are consistent with the invariance of S. The current state of relativistic statistical mechanics and thermodynamics is, in actuality, one of rather wide disagreement on the question of what are the transformation characteristics of the various thermodynamic variables.[3] But there is general agreement that entropy should be invariant; perhaps the best basis for this invariance is that entropy is a measure of the configuration or order of the elements of a system (see sec. 4, Chap. IV), and the order would not be expected to change as a result of motion of the observer relative to the system. In the words of R. C. Tolman,[4] ". . . this result (that $S=S'$) is in agreement with the statistical mechanical interpretation of entropy in terms of probability, since the probability of finding a system in a given state should evidently be independent of the velocity of the observer relative to it."

The argument for invariance of entropy is particularly compelling for a system which is in equilibrium and whose entropy is therefore constant in time. If the entropy is changing in time, *and* the system is of non-negligible spatial extension, a theorem of simple entropy invariance for the entire system is not always adequate. Suppose that volume elements V_1 and V_2 of the system G, at x_1 and x_2 in Σ, have entropies s_1 and s_2 respectively at time t, but slightly different entropies at time $t+\delta t$. Then, for an observer in the relatively moving system Σ', the spatial separation $x_2-x_1=\delta x$ has a time interval $\delta t=\delta xv/c^2$ associated with it. Therefore, the Σ' observer, if in physical interaction with the Σ system, must expect to find an effective time interval between V_1 and V_2: s_1 and s_2 for him are observed at t and $t+\delta t$ respectively, if they are both at t for the Σ observer. The entropy invariance is, therefore, only for an element that is not extended in the direction of relative

[4] *op. cit.*, Reference 3, p. 158.

motion, if the entropy is changing in time. In our further use of entropy we shall not require consideration of any other relation than that of simple invariance.

Let us now assume that G is a thermodynamic system for which the entropy may be determined and which will serve as a clock: again, our gas clock may be taken as an example. The entropy S is then a function of t, such that at t_1 the entropy of G is S_1, at t_2 it is S_2, etc. We require G to be of negligible extension along the x direction, so that the problem of a variation of t' with $\delta x'$ does not enter when G is observed from the relatively moving coordinate system. (Or, alternatively, we could stop the clock G for the required time δt during which time measurements are made from Σ and Σ', so as to avoid the problem of time variation with spatial extension; that is, we could require that G be stopped at t during the interval δt which for Σ' observers is associated with a spatial extension.)

The entropy invariance theorem may be taken in the following form, given by R. C. Tolman: ". . . we shall add to our thermodynamic requirements for systems in a state of rest or uniform motion, the requirement that the entropy of a system would be unaltered by reversible adiabatic change in velocity without absorption of heat."[4a] With acceptance of this theorem, we have a strongly relevant consideration in the discussion of the problems that arise with respect to relativistic changes of clock rate with velocity. We note, first, that entropy invariance would allow us to present clock paradox situations in a manner that permits no change of clock-time as a result of acceleration. If the clock in question is a thermodynamic system, like our gas system G, and it is thought of as moving away from a stationary clock for a period of time T at a uniform velocity, and then having its velocity reversed and returning to the stationary clock, we can rule out the possibility of there being significant change in the time of G as a result of its undergoing velocity reversal. The period of acceleration may be made small compared with T, and the invariance of entropy tells us that acceleration to another coordinate system will not in itself have any effect on the entropy, and hence the time-reading, of G. Likewise, the time of the "stationary" clock will not change, from the point of view of the G system, as a result of the change in motion of G.

It might be objected that the preceding argument is invalid because the time of a clock might be a different function of its entropy in one coordinate system than in another. It can readily be shown, however, that such a difference would involve a conflict with the principle of relativity. Suppose we have two similarly constructed clocks, G, for which

[4a] *op. cit.*, Reference 3, p. 157. Virtually identical statements are made by Pauli, *op. cit.*, Ref. 3, and by C. Moller in his *Theory of Relativity*, Oxford University Press, 1952, p. 212.

the time t is some function of the entropy, $t=f(S)$, when the clocks are calibrated in the same inertial system. If we argue that if one of the clocks is placed in a relatively moving inertial system a new time calibration, in that system, will give a new functional relation, $t'=f'(S)$, we have stated that measurements made within an inertial system will give a determination of the state of motion of the system; for, the function f relating time-reading and entropy of the clock would differ from system to system. The principle of relativity explicitly requires, however, that the state of uniform motion of an inertial system have no discernible physical effects within the system.

We can further use the entropy invariance theorem as a basis for arguing against there being any kind of change in time-rate of a thermodynamic clock as a result of relative motion to an observer. Let us now assume that we have an indefinite number of gas clocks G set out along the x-axis of coordinate system Σ, and one additional clock, Σ', similar in construction to the others, which is initially also at rest in Σ. All of the clocks are synchronized in Σ, and all are continuously gaining entropy increments δS. The clock Σ' is now accelerated through velocity increments δv, by a process that in the limit is adiabatic and reversible. In accordance with the entropy invariance theorem, the entropy of Σ' at any velocity v will be independent of all previous velocity increments; and further, the entropy increment δS at any velocity is independent of velocity, in accordance with the theorem. In symbols, we may write the invariance theorem as $d/dv(\delta S)=0$. But if S were any function g of v, $S=g(v)$, we would not have $d/dv(\delta S)=0$. We must take it, then, that the Σ' clock while being accelerated gains the same increments δS which comparable Σ clocks are gaining; if it were otherwise, entropy would not be independent of velocity. In the limiting case of zero velocity increments, we must also have the same entropy increments for the Σ and Σ' clocks, and hence also the same increases in clock readings. We conclude that similar entropy clocks, in relative uniform motion, will run at the same rate.[5]

The heart of the preceding argument is the passage from entropy invariance under a relative velocity change to invariance of *rate* of entropy-increase under such a change. We argue that if entropy is indifferent to relative velocity for an observer, i.e., if $d/dv(\delta S)=0$, then there can be no difference in the entropy increments that comparable thermodynamic systems gain, as a result of relative motion. The argument seems to be a sound one, to the extent that one considers physical processes as being prior in nature to time, so that time measure

[5]The contradiction between entropy invariance and relativistic time changes of clocks was the basis for the initial suggestion of the distinction between Lorentz and Clausius processes; R. Schlegel (Abstr.), *Physical Rev.*, *87*, 198 (1952).

is determined by these processes. If one takes the view, however, that time rate somehow determines physical processes, in a way that transcends ordinary physical theory and observation, one might hold that the independence of entropy change to velocity is of no consequence, and that the rates of entropy increase in two thermodynamic systems must differ as a result of relative velocity between the systems.

5. RELATIVE AND ABSOLUTE TIME.

On being asked to explain in one sentence the basic tenet of the theory of relativity, Albert Einstein is reported on an occasion to have replied as follows:[6]

> "If you don't take my words too seriously, I would say this: If we assume that all matter would disappear from the world, then, before relativity, one believed that space and time would continue existing in an empty world. But, according to the theory of relativity, if matter and its motion disappeared there would no longer be any space or time."

Our basic approach in this book is in that same tradition, of time's being a property of physical processes, in which Einstein places relativity theory. This view of time was held by Aristotle, who wrote[7] that time cannot be separated from movement, and is "the calculable measure or dimension of motion with respect to before-and-afterness," and was clearly expressed late in the Greco-Roman period by St. Augustine.[8] Under the influence of Newton, the idea that time is "in and of itself" and "everywhere the same" came to dominance, at least in physics. It was one of Einstein's greatest achievements to have brought physics into the tradition that time is *not* prior to and independent of the world of physical events.

An even more revolutionary aspect of Einstein's special theory of relativity was its statement of the variation of time rate with relative motion. We have seen, in Chapter VII, that this variation has been confirmed beyond doubt in physical observations. The frequencies of light and matter are changed as a result of relative motion, the life-times of decaying particles are changed, and the relativity of simultaneity is directly observed in de Broglie waves. In all of these instances, the process undergoing relativistic time change is involved in a physical interaction at high relative velocity with the observing system. That is, the changes are for Lorentz processes. For such processes, the application of the relativistic time transformation must be regarded as a firm keystone in our physical knowledge.

---

[6]Quoted by Philipp Frank in *Philosophy of Science,* Prentice-Hall, New York, 1957, p. 123.

[7]*Physica,* IV (Loeb Classical Library).

[8]See, e.g., Bertrand Russell, *History of Western Philosophy,* George Allen and Unwin, Ltd., London, 1946, p. 373.

We have suggested, however, that there is another class of processes, those which we have called Clausius processes, for which the relativistic transformations are not applicable, even though one of these processes is at a high relative velocity to the observer; in such a process, there is no physical interaction with the observer that is an essential part of the process. Since the Clausius processes are a group whose time rates are invariant, they may be used to define a common, universal time. Specifically, as clock processes they would have time rates that are invariant to relative uniform motion; and further, because of the invariance of light speed *c*, observers in various parts of the universe could use light signals to synchronize their clocks to a common time scale, even though they were in relative uniform motion.

The invariance of Clausius processes enables us to accept the absolute time of Newton—time that is "everywhere the same"—as physically existent, and yet also, we have in Lorentz processes the physical variation of time rates with motion that is required by relativity theory. We have certainly no reason to consider the time of Clausius processes to be absolute in the sense of being prior to physical process, but we can consider this time to be uniform to the same extent that we ascribe uniformity to physical processes generally. That is, having justified the independence of clocks from any effects of their uniform and relative motions, we have to consider whether physical processes may otherwise vary in rate from time to time or place to place in the universe. We generally feel confident that such variation does not occur—except perhaps over very long periods of time, as discussed in section 3, Chapter I—and we call this lack of variation the "uniformity of nature." The Newtonian property of time's being "everywhere the same" is then, in our view, a property of time rates of Clausius processes that has as much validity as does our assumption of the uniformity of nature.

When we refer to the "time" of an event we usually mean the time as given by a clock, that is, by a Clausius process. The invariance of these processes to motion, if correct, does therefore restore a universal physical time. The (transformable) time of Lorentz processes is likewise usually measured in terms of clocks that are Clausius processes; for example, we measure the time of meson decay, or the frequency of light from a rapidly moving source, in terms of the laboratory (Clausius) clock. The Clausius processes, then, provide a universal and (at least approximately) unchanging temporal framework against which the relativistically transformed time of Lorentz processes may be measured.

The time invariance of Clausius processes can play an important role in cosmological theory. One of the long-standing problems in that theory is, How is a meaningful time scale to be defined for cosmological processes, in the face of the relativistic time transformation? For, if the usual special relativity transformation is applied, the time of the evolution of a given galaxy, for example, will depend upon the relative motion of the observer with respect to the galaxy. And, we know that relative galactic motions are at least as much as one-fifth of the speed of light *c* (see sec. 4, Chap. V). Yet, the time of cosmological theory is presumably a "cosmic time," which has significance throughout the universe and not only for the observer in, say, the Milky Way galaxy. In the words of H. Bondi: "This difficulty (of cosmic time) is common to all (cosmological) theories . . . , but unfortunately not very much attention seems to have been paid to resolving this problem so vital for the logical foundations of all these theories."[9]

A common procedure in cosmological theory, with respect to the cosmic time problem, has been to adopt an assumption first stated by H. Weyl, and known as "Weyl's postulate."[10] According to this postulate, the various parts of the universe—the various galaxies we might say, for definiteness—are not unrelated. Rather, the galaxies had a common origin at some time indefinitely long ago. The common development of the galaxies therefore carries with it a time variable which at any point may be given a value that is common to all the galaxies. In more technical language, the world-lines (mapped paths) of the galaxies in a suitably chosen four-dimensional space-time form a bundle which diverges from some common point; a surface normal to all the lines is everywhere at a common time, with the length of the lines from origin to surface being a measure of that time value.

With the cosmological model which depicts the present universe of galaxies as resulting from an expansion of some original condensed state, Weyl's postulate is justifiable. There are other models, as for example the steady-state one, to which it is not applicable. In any event, his postulate solves the problem of the incompatibility of relativistic time changes with a universal cosmic time in an artificial and *ad hoc* manner. A basis that is firmly rooted in the structure of physics is desirable if we are to believe with conviction that there is a cosmic scale that is significant beyond the frame of reference of any one observer.

The non-applicability of Lorentz transformations to Clausius processes provides such a justification for a cosmic time scale. A given galactic process is without essential interaction with an observer who is physically removed from the galaxy; the velocity relative to the observer is not, therefore, a factor of relevance in the time rate which the observer will find for a given galactic process. Again we emphasize

[9]H. Bondi, *Cosmology,* Cambridge Univ. Press, 1952, p. 72.

[10]See "Relativistic Cosmology," by H. P. Robertson, *Rev. Modern Physics, 5,* 62 (1933), and discussion by H. Bondi, *op. cit.,* sec 10.5 and p. 150ff. Weyl's discussions are in *Physik. Zeits., 24,* 230 (1923) and *Phil. Mag., 9,* 936 (1930).

that any process of the galaxy which does involve physical interaction with the observer will be subject to a Lorentz transformation: thus, certainly we expect the relativistic Doppler effect to be valid for galactic radiation which reaches an observer moving at high relative velocity. But the internal galactic (Clausius) processes go at a rate which is determined by the nature of those processes, and not by the state of uniform motion of the galaxy.

A given kind of process, e.g., the nuclear burning of a mass of hydrogen atoms, would therefore occur at the same rate in any coordinate system; and, a given process used internally as a rest clock would have the same rate in any uniformly moving coordinate system. Observers anywhere in the universe, regardless of their state of uniform motion, would therefore judge processes by the same time scale, using rest clocks, and all physical processes that are physically similar would occur with the same time rate, regardless of the uniform motion of the galaxy in which they were occurring. Studies of process throughout the observable universe would therefore give observers everywhere a common time scale. But the profound lesson that we have learned from relativity theory is that there are also in nature an abundance of physical interactions (Lorentz processes) in which changes in time rate do occur as a result of high relative uniform motion.

6. SUMMARY AND DISCUSSION.

We have seen that with our proposed distinction between Lorentz and Clausius processes we readily solve the "clock problems" of relativity theory, since there would then be no transformation of "public," non-interacting clock mechanisms. On the other hand, meson decay times would be relativistically transformed whenever the relativistic increase of meson life-time results from relative motion between the meson and the matter with which it interacts (see sec. 7, Chap. VII). The entropy invariance of a clock-like system (sec. 4, Chap. VIII) would be brought into complete harmony with what is expected as a result of Lorentz time-transformations. And, with the distinction, constancy of light speed c (sec. 2, Chap. VIII) may be ensured directly as a result of Lorentz transformation of photons in physical interaction with an observer, with predicted changes in photon properties as observed in the Doppler effect phenomena; the assumption of constant c only as a result of cumbersome and physically unconfirmed relativistic changes in all clocks and rods of a system may then be dropped.

An argument of a general nature in favor of the distinction between Lorentz and Clausius processes rests on the central role of physical interactions in the Lorentz processes. The relativistic time changes that are observed occur only in connection with physical interactions between the process which undergoes the transformation and the observing apparatus. That is, the transformations occur in definite physical situations, with physical fields or particles. The restriction of relativistic changes to situations involving physical interactions appears therefore to be a natural one. If relativistic time changes occur also for those processes which we have denoted as Clausius processes, then relativistic time changes are occurring in association with situations of relative motion which do not involve an interaction of the kind that we have come to observe and understand as a physical interaction.

It is unsatisfactory indeed to the physicist, to require that there be purely internal transformations within a physical system without any physical interaction which might serve as a basis for the transformation. A transformation which appears without reference to any physical interaction has no physical aspects which a physicist can study in hopes of gaining understanding of the nature of the transformation. In contrast to hypothetical clocks, actual relativistic transformations (Lorentz processes) are physical processes which can be studied as such. In the some fifty years that have passed since the first introduction of special relativity theory, there has been no advance in the basic physical understanding of the Lorentz transformations, and during this period the transformations have generally been taken simply as a result of relative motion. It seems likely that if the transformations were studied as physical interactions, progress might well be made in the understanding of them.

We might, for example, ask, What are those properties of electromagnetic radiation which, in a photon-observer interaction, require the selection of the wave-length and frequency that give a relative velocity c? Or, what is the relation between the increased energy that decay products of a relativistic meson must bring to the matter with which they interact, and the increase in life-time of the meson? There is an obvious difference between the interactions that give rise to the Doppler effect for light and those that give the observed relativistic meson life-time changes. The coupling between an observer and electromagnetic radiation must be such that in the interaction there is a selection, from among the continuum of frequencies that is potentially available, of that frequency which gives the velocity c. The source has determined the frequency for an observer at a given velocity relative to it, but there is no indication that the source has any other role in the observer-radiation interaction. In the meson case, in contrast, the observer-source coupling must be quite different; the decay products of the meson interact with the particles of the matter through which the meson is moving, and the meson life-time is related to that relative velocity. Hence, the relative velocity of the decay products, and therefore the energy which they bring to their interactions, must in some maner determine the source life-time.

In a few words we may say, then, that with the distinction between Lorentz and Clausius processes we assure the constant value of light velocity for any observer as a result of properties of the electromagnetic radiation which interacts with a given observer, and we find the many relativistic time transformations of nature to be associated with an interaction at high relative velocity. By removing the hypothetical changes in clock rates in situations that do not involve interactions at relativistic velocities, we bring the theory of relativistic time changes into the domain of physical interactions, and we obtain complete consistency among the various relativistic time changes and with relativistic thermodynamics. The distinction retains the many confirmed applications of the Lorentz transformations to relativistic physical interactions, and keeps for relativity theory the same great utility as a tool in physics investigations that it has at present, because the concern of physicists is with processes of physical interaction. It seems to me, therefore, that the distinction may very well be a valid one. On the other hand, it is only fair to emphasize that the traditional "clock-rate change" interpretation has to date had many successes in its guidance of physical thought and experiment. It seems reasonable to ask, therefore, not only if there may be experimental evidence that confirms the proposed new distinction, but also, whether or not there may be further support which will maintain the traditional doctrine of relativistic time changes for all clock processes.

We shall first consider the question of experimental confirmation. There is of course a large body of evidence which establishes the existence of relativistic transformations for Lorentz processes. There is no experimental evidence for time (or other relativistic) transformations in what we have termed Clausius processes, and we have in this fact a kind of negative empirical support for the distinction between Clausius and Lorentz processes. Nonetheless, one would hesitate to accept our proposed difference between the two kinds of processes until an empirical test had verified the proposition that for Clausius processes time rates are independent of state of uniform motion. Experimental techniques which are currently available do not permit, at least in any obvious way, of such a test, but perhaps the day is not far off when a test will be possible. One method, which is simple in conception and which seems almost to be within the reach of present-day experimental physics, would be to accelerate to a high velocity molecular clusters of atoms in which one of the atoms was radio-active, with a short half-life. If the cluster could be maintained at the high velocity for a period of time, and if a selection could be made of those clusters in which the decay product (perhaps an α-particle) of the radio-active atom interacted with one of the atoms of the cluster, we would have a test of the proposed invariance of Clausius processes. The decay of a radio-active atom with respect to other atoms associated with it in a large molecule would be a physical interaction in which no relativistic velocities were involved. Hence, even though the entire cluster had a relativistic velocity with respect to the laboratory, if the Lorentz-Clausius distinction is valid there should be no effect on the life-time of the radio-active atom. Contrariwise, with the "clock-rate change" doctrine of application of Lorentz equations to all processes, the radio-active atoms should show a transformed life-time simply by virtue of their motion relative to the laboratory. There would then be no difference in life-time of a nucleus, whether its decay products interacted with other (relatively stationary) atoms in its cluster or with (relatively moving) laboratory atoms.

Another way of testing the relativistic transformation of a Clausius process would be through studies of the rate of a clock moving in an artificial earth satellite. Efforts towards such an experiment are under way, and it appears possible that by about 1970 first results will have been obtained. The experiment, which was discussed with respect to the relativity theory involved in 1956 by S. F. Singer,[11] involves sending a clock into orbit, with apparatus which counts pulses of the clock and periodically sends information about its count to the ground, where comparison is made with the number of pulses of a similar earth-bound clock. The atomic clocks now in operation, with accuracies of the order of one part in 10^{10}, would easily have sufficient precision for any effects to be expected when the clock has the relative velocity of an earth satellite. Such a clock is a Clausius process. The essential clock mechanism involves applied oscillating electromagnetic fields, and transitions in atoms between energy states that arise from interactions between the atomic nuclei and their extra-nuclear electrons. The atoms serve as a clock by virtue of the fact that the energy transitions will occur only when the oscillating field which induces the transition has the required characteristic frequency; by suitable responses of the atoms the frequency of the field-producing apparatus may be corrected when it varies from that characteristic frequency. A counter, then, which is in the satellite with the clock and which is counting the oscillations of the field-producing apparatus would not be

[11]*Physical Review, 104,* 11 (1956). In the studies of radiation frequency-shifts between axis and rim of a rotor (H. J. Hay, J. P. Schiffer, T. E. Cranshaw, and P. A. Egelstaff, *Phys. Rev. Letters, 4,* 165 (1960); W. Kündig, *Phys. Rev., 129,* 2371 (1963)) a Lorentz process, specifically, the transverse Doppler shift, is clearly involved. Likewise, the Pound-Rebka experiment (R. V. Pound and J. L. Snider, *Phys. Rev., 140,* B788 (1965)), in which gamma radiation interacts with a gravitational field, is not a test of a Clausius process. In general, for a Clausius process, in contrast to a Lorentz process, there is not an energy-transfer corresponding to the fourth component of the momentum vector derived from that space-time vector which entails the time transformation.

in relativistic interaction with the clock, even though the satellite were moving at relativistic velocities with respect to the earth. If the pulses of the atomic clock in the satellite were decreased in rate as a result of relative motion, we would have a clear vindication of the "clock-rate change" interpretation of relativity theory. If there were no rate decrease, something like our proposed Clausius-Lorentz distinction would merit further consideration.

There actually are two different temporal effects to be considered in the satellite-clock experiment. If we take the usual "clock-rate change" interpretation, the relative motion with respect to the earth should give the clock-slowing that is predicted by the special relativity theory; but also, the difference in gravitational potential between a point on the earth's surface and a point above the surface should entail, in accordance with the general theory of relativity, a difference in time-rate for physical processes at the two points (see the discussion in sec. 4 of Chap. VII). At low gravitational potentials (i.e., when the clock is close to the earth's surface) the clock rate should be slower than at higher potentials. The velocity and altitude of an earth-orbiting satellite are, of course, in a fixed relation to each other, and it therefore is possible to assign relative magnitudes of the two effects at different heights. The clock should show only the special relativity theory slowing-down, compared with an earth clock, when it is at very low altitudes, while the general relativity theory increase in clock rate becomes dominant when the clock is at high altitudes, where the gravitational potential is much increased over the (negative) value it has on the earth's surface. The two effects balance each other at an altitude of one-half of the earth's radius (about 2000 miles), and the satellite-clock at that altitude would be predicted to have the same rate as an earth clock. Experiment of course may show the physical existence of only one, or both, or neither of the two effects.

If the earth-satellite experiment is to give a test of the Clausius-Lorentz process distinction, it is important that the satellite clock pulses are counted in the satellite, rather than transmitted radio-wise to an earth counter. In the latter case, the impulses would be subject to the relativistic Doppler effect, and it would not be clear whether an observed decrease in clock rate were the result of an actual decrease in that rate or the result of the Doppler shift on the radio signals from the moving satellite. The Ives-Stilwell experiment with rapidly moving H atoms (see sec. 3, Chap. VI) has confirmed the existence of that Doppler shift, but also, this experiment has been taken as a confirmation of the relativistic slowing down of the atomic processes in the emitting H atoms.[12] The atoms, functioning as tiny clocks, are inter-

[12]See, e.g., H. P. Robertson, "Postulate *versus* Observation in the Special Theory of Relativity," *Rev. Modern Physics, 21,* 378 (1949).

preted as being slower in their internal processes by the factor $\sqrt{1-\beta^2}$, where v is the velocity of the atom with respect to the observer; as a result of this rate decrease they give the observed relativistic shift, in their emitted light, to frequencies that are smaller by the $\sqrt{1-\beta^2}$ factor. The experiment may however be regarded as supporting either the "clock-rate change" interpretation or our proposed Lorentz-Clausius distinction, and we shall now discuss it as a final illustrative comparison between these two points of view.

We have seen that the Doppler equation is a consequence of applying the Lorentz transformation (from a system S to a system S') to electromagnetic radiation. The equation holds for phenomena moving with the velocity of light, and has no reference whatsoever to the source of the radiation. Physically, we can emphasize this point by thinking of light from a distant star; the star may have ceased to exist by the time its light reaches the earth, but we have no doubt that high relative velocities set up in a terrestrial laboratory would give relativistic Doppler effects for the observed starlight. On the "clock-rate change" view, the effect exists because the electromagnetic radiation appears differently to S and S' observers; clock and time rates are different in those two systems, and the radiation, unchanged "in itself," is received by observers or sets of apparatus that have different time rates, even though identical if at rest in the same system. If the atoms or other radiation generators were in existence, they too would appear to have different frequencies when compared with rest clocks in systems S and S' respectively. The radiation and source frequency would each differ by the same factor for S observers, as compared with S' observers, because, to repeat, the source of the difference is the difference in S and S' clock rates.

On our suggested "interaction" point of view, the internal mechanisms of the radiation source would be Clausius processes that are unaffected in rate by uniform motion. The relativistic Doppler shift would be an interaction effect, associated with relative differences in speed between S and S' observers. The photon frequencies would be considered to be literally and physically different for the two observers, in that the photon frequency which is permitted (or selected) for interaction with an S observer would be different from the frequency for interaction with an S' observer: in each case, natural process operates so as to give a frequency that will maintain constant light speed c.

Both of the alternative points of view are consistent with the Ives-Stilwell results. Those who hold the "clock-change rate" doctrine must be wary, however, of bringing in any aspect of the interaction point of view. If one holds that atomic clock-processes run slower as a result of being in a relatively moving system, and *also* thinks of the relativistic Doppler effect as being a physical transformation of electromagnetic waves, he is then faced with the following untenable conclusion. On the clock-rate change point of view, the source of electromagnetic waves runs slower in its internal processes by a factor $\sqrt{1-\beta^2}$, as a result of relative motion at velocity v. For vividness one might think of a large-scale generator that is moving at relative velocity v and emitting radiation of the same frequency as its rotational frequency; the emitted waves are then decreased in frequency by the $\sqrt{1-\beta^2}$ factor. But if the electromagnetic radiation is itself subject to the relativistic Doppler shift, we thereby have another $\sqrt{1-\beta^2}$ factor. A total shift by $(1-\beta^2)$ rather than by the observed $\sqrt{1-\beta^2}$ factor would then be expected. In maintaining the "clock-rate change" doctrine we must therefore consistently consider that relativistic changes in processes of some one coordinate system are consequences of time-rate differences between that system and the observers' coordinate systems. On our proposed interaction view, these same changes are to be taken as physical changes in the processes of interaction between observers at different relative velocities; for, on this view, similar rest clocks run at the same rate in all (non-accelerated) coordinate systems.

If valid, the interaction point of view and its distinction between Lorentz and Clausius processes does solve the "clock problems" of special relativity theory. But we now look again into what is required if one is to solve these problems and still retain the "clock-rate change" interpretation. We saw that the essential element in these solutions is the role of a preferred coordinate system. Two kinds of "preference" may be distinguished: 1) the setting-up of any inertial system as a preferred coordinate system; 2) the assumption of a cosmic coordinate system, usually related to the distribution of matter in the universe.

With the first kind of preferred system we clearly are on firm ground, in that nature does present us with inertial systems: systems in which the equations of physics as written for situations with no physical forces present are found to be valid. On the other hand, it is not certain in the clock problems that the fact that one clock is always in the same inertial system, while other clocks change from one system to another, is sufficient basis for a solution. One clearly may calculate a completely consistent solution on the basis of the time readings of the always-an-inertial system; but also one may calculate a consistent but different solution on the basis of a clock which changes from one inertial system to another. If one regards it as legitimate to neglect the effect of the change from one system to another, and if one judges that the total physical situation is not necessarily asymmetrically controlled by the times as calculated from the continuous inertial system, then the existence of inertial systems does not lead to an unambiguous solution of the "clock problems."

With assumption of the second kind of preferred system—one which in some way defines state of motion or rest with respect to the universe—it is, as we have seen, easy to solve the twin or rocket paradox; for then it is possible to pick out one system as being the moving system, in which the special relativity time effects are preferentially existent. The idea of such a preferred coordinate system has some qualitative physical basis in what is known as Mach's Principle,[13] which states that the structure of space and hence the inertial properties of matter are determined by the matter in space. The inertia of matter—or, the existence of three independent spatial directions for which conservation of momentum is valid—is then in some way intimately associated with the total amount and disposition of matter in space. The Principle is only an hypothesis, and it has not been expressed in detail in physical theory or, needless to say, confirmed in observation. Nonetheless, strongly appealing arguments can be given for it. It is an easy step from an acceptance of Mach's Principle to the idea that the galactic universe determines a coordinate system, with respect to which motion may be defined. The earth would presumably be—compared with the very high speeds which are required for relativity theory effects—more or less at rest with respect to this system. Rockets or mesons or electrons moving with speeds approaching the speed of light with respect to the earth would then be experiencing relativistic effects that were not symmetrical; that is, similar effects would not be expected on the earth, from the point of view of the moving systems. The relativistic time and space changes would ensure the constancy of light speed c, and explain the null result of experiments, like the Michelson-Morley experiment, which were designed to show the existence of a preferred coordinate system.

It must be pointed out, however, that the appeal to the universe of stars or galaxies for a fixed reference frame is a highly ambiguous one. The relatively small random motion of the near galaxies among themselves may be overlooked, but the farthest observed galaxies are receding from us with velocities of the order of one-fifth the speed of light, and almost certainly the speed is greater for farther outlying matter. Cosmological theory, as we have discussed, indicates that these recessional motions are a result of a general expansion, so that the observer in any given galaxy finds himself to be at a center of receding galaxies. It might be necessary to maintain, then, that there is not one preferred frame in the universe, but rather a frame roughly associated with a large spatial domain of each galaxy or

[13]Ernst Mach, *The Science of Mechanics*, trans. by Thos. J. McCormack, Open Court Publ. Co., LaSalle, Illinois, 1942, pp. 285-290.

group of galaxies. All this, however, is highly speculative, even from the vantage point of cosmological theory.

With the acceptance of a preferred cosmic coordinate system, we resolve the clock paradox problem. Also, there would then be no problems like the one of an increased meson life-time by clocks both in the meson and the laboratory-counter coordinate systems; for there would be no reciprocity of relativistic effects between the meson and laboratory (earth) systems. However, the reconciliation of entropy invariance with change in clock rate is not helped by assumption of a preferred system. Also, it is not to be forgotten that such an assumption brings a profound modification to special relativity theory; for the assumption contradicts the principle that between any two relatively moving systems the effects of motion are the same, from either system to the other, or, equivalently, that there are no effects of uniform motion within a system. This basic relativity principle has so far been a remarkably fruitful and well-confirmed postulate in physical theory. It has been widely utilized in atomic and field-theory physics, and many far-reaching theoretical changes would be required if the principle were dropped. Quite possibly, if intensive consideration were given to the denial of the principle, additional specific objections to such a denial would be found.

Our final conclusion with respect to the problem of internal relativistic time changes in complex systems must be that the matter will only be settled by appeal to experiment. The relativistic time changes that are found in nature, in association with physical interactions, give a satisfying conviction of the validity and insight of special relativity theory. But in spite of the positive statements that various writers have made with respect to relativistic changes in time rates in large-scale clocks and biological organisms, we have seen that there are arguments against acceptance of these changes. If observation should find such rate changes not to occur, the interaction hypothesis presented in this chapter would give a way of reconciling that result with the theory of relativity. If the changes are found to occur (in accordance with what is perhaps the expectation of most physicists) the solution of the relativistic clock problems in terms of the role of inertial or preferred coordinate systems can hardly be questioned; and, the foundations will then be greatly strengthened for further study of the role of these systems in natural processes.

Chapter IX

Time and Quantum Theory

1. INTRODUCTION.

Quantum theory and relativity theory are the two great sets of theoretical concepts which have been introduced into physics during the twentieth century. We have discussed the far-reaching effects of relativity theory on our concept of time: the basis of time in physical events was emphasized by Einstein, and previous notions of time were liberalized and broadened by his realization of the effect of high relative velocity on time rates. The concepts of quantum theory, in contrast, have not so much changed our primary ideas about time. Nonetheless, quantum physics has obvious relevance to our physical time concept, and in this chapter we shall see some ways in which quantum theory gives us important information about physical time.

There are two basic innovations among the physical ideas of quantum theory. The first of these was contributed by Max Planck in 1900 in connection with the problem of deducing a satisfactory equation for the variation of energy-density with wave-length in the radiation emitted by a hot body.

It is a consequence of Maxwell's equations for the electromagnetic field that a charged body will emit radiation if the body is accelerated. The equations do not give any information, however, about the wave-length distribution of the emitted radiation. Shortly before the time of Planck's work, it had been found by Lord Rayleigh (J. W. Strutt) that the use of statistical mechanical concepts and the assumption of energy radiation by accelerated, charged atomic bodies in matter would give a partially correct equation for the energy distribution: the agreement with observation was satisfactory for long wave-length radiation, but altogether wrong for the shorter wave-length energy in the radiation from a given body. Planck found that he could derive a correct equation for the frequency distribution of radiant energy if he made an important change in the assumptions used by Rayleigh; instead of assuming that radiation was continuously emitted (or absorbed) by accelerated, charged particles in matter, he proposed that radiation is exchanged between an atom and a "radiation field" only in packets or quanta. The magnitudes of the quanta are to be proportional to the frequency ν of the radiation. That is, Planck's new assumption was that radiant energy is in quanta of energy $E=h\nu$, where h is the proportionality constant that we now call Planck's constant.

This proposal was both daring and unexpected, for there had been nothing in the preceding development of radiation physics to suggest a "quantization" of radiant energy. The success of the new assumption in leading to a correct description of the energy-frequency relation of radiant energy could not, however, be matched by any other theoretical proposal. And, further evidence of the validity of the quantization was soon forthcoming. In 1905, Einstein gave physics the first satisfactory equation for the photo-electric effect, the phenomenon in which incident light ejects electrons from certain metallic surfaces, and he did this by directly applying the Planck concept of energy $h\nu$ for photons. A year later Einstein used the quantum concept to explain anomalies in the specific heats of metals at low temperatures. In 1913 Niels Bohr, in one of the great triumphs of modern physics, showed that Planck's energy quantization, taken with the nuclear model of the atom, led to an equation that very accurately predicts the wave-lengths of the radiation emitted by excited hydrogen atoms.

There is by now an overwhelming abundance of evidence that confirms the $E=h\nu$ equation. It is to be emphasized that there is no disposition on the part of physicists to interpret the equation as being primarily a violation of the Maxwell electromagnetic equations which predict continuous radiation. For it is found that for large-scale electrical phenomena—of the kind, for example, that is of concern to the electrical engineer—the radiation of accelerated charges is as predicted by the electromagnetic field equations: thus, the equations correctly describe the relations between electron oscillations and radiation emission in the antenna of a radio transmitter. But when individual atomic radiations must be considered in theory or practice, the quantum nature of radiation may become important. It is, then, the quantum nature of radiation which puts a limit on the applicability of the Maxwell equations for electromagnetic radiation. Those equations are correct as a smoothed-over description of radiation processes that involve many quanta; but in considering detailed processes of radiation emission and absorption, we must take account of the fact that the radiation of frequency ν is in energy packets $h\nu$. The magnitude of h—6.6×10^{-27} erg-seconds—shows us that by every-day standards the individual quanta of energy are small for the kind of radiation that we ordinarily experience: for visible light ν is about 10^{15} sec^{-1}, and even for very energetic x-rays ν is only about 10^{20} sec^{-1}.

The second innovation that is basic to quantum theory is the conception of the wave nature of matter, proposed by Louis de Broglie in 1923. We have discussed de Broglie waves at length in section 5 of Chapter VI, but we here point out that there is a significant element of symmetry between the two innovations. The quantization of radiant energy gives a corpuscular aspect to radiation, since the quanta are definite in magnitude of energy, and also since it is found that in interactions in which the quanta give up their energy and momentum they effectively do so at a point, like a corpuscle. The particles of matter, on the other hand, are shown by the de Broglie equation to have wave properties: the particle is not a simple corpuscle, but rather, as we have seen, it is spread out in a wave form when it is observed at a high relative velocity.

The introduction of the wave property of matter provided a key for the solution of some difficulties that had arisen in the first development of quantum physics; with these corrections the physics of the relations between atomic nuclei and the electrons of atoms and molecules became, as far as basic principles are concerned, very nearly a completed subject. And further, with the wave conception, it has been possible to develop adequate theoretical descriptions of the interactions between the elementary particles of nature, insofar as those interactions are dependent on the wave properties of matter. And likewise the interaction properties of photons with matter have been thoroughly developed within the framework of quantum physics. The center of interest in fundamental quantum theory has, by the year of this writing, come to be the problem of quantization itself: the problem of how can we understand the definiteness in magnitude of the elementary units of energy and electrical charge that nature presents to us.

2. THE UNCERTAINTY PRINCIPLE.

The manner in which the wave and corpuscular aspects of radiation and matter are brought together in quantum theory makes a long and not very simple story, and we shall not attempt to tell it in this book. It is because of these two aspects of nature, however, that physicists have been led to accept a principle that is of significance for our study of physical time: this is the "uncertainty principle" of quantum theory. The principle was first enunciated by Werner Heisenberg in 1927 in connection with considerations of the role of quanta in physical measurements, and has been generalized by others into a basic feature of the entire quantum theory. The discussion of the principle which follows in this section is largely based on that given in a paper by Bohr.[1] In this treatment the close association between physical uncertainty and wave-particle properties is clearly expressed.

We shall consider a wave of frequency ν and wave-lengths λ_x, λ_y, and λ_z in the x, y, and z directions respectively. For convenience we express the wave-lengths as wave-numbers k_x, k_y, k_z, which are the reciprocals of the wave-lengths; for example wave-number $k_x=1/\lambda_x$. With these symbols, a simple plane wave in three dimensions can be represented by $A \cos 2\pi\ (\nu t - xk_x - yk_y - zk_z + \delta)$, where A is an amplitude and δ an arbitrary phase factor.

We now ask how we may superimpose waves of varying ν and k values, so that we get as a result of the superposition a non-vanishing wave field in a limited region only. That is, we consider that each wave originally extends indefinitely. We know that on superposition we can cancel or re-enforce waves: thus, two waves of equal ν and k's which are out of phase by a half wave-length will everywhere "destructively" interfere with each other since each positive amplitude component of one wave will be annihilated by a negative component of the other; but the same two waves, if in phase with each other, will "constructively" re-enforce each other at every point. We want the more complicated situation, however, in which a group of waves so interfere with each other that we have net non-zero wave amplitudes in a given spatial region, but outside this region the waves, having completely interfered with each other, give a net zero amplitude.

It is a result of the theory of waves—we shall not give the derivation—that in order absolutely to achieve the condition of non-vanishing amplitude in a limited region only, one must have an infinite number of waves, corresponding to all values of ν and k_x, k_y, k_z. Further, it can be shown that there is a relation between the space and time extension of the region in which the waves do not disappear and the mean difference between the values of ν or of k for all pairs of elementary waves in that region. The relation may be given in the following approximate but simple form:

$$\Delta t \cdot \Delta \nu = \Delta x \cdot \Delta k_x = \Delta y \cdot \Delta k_y = \Delta z \cdot \Delta k_z = 1. \qquad (1)$$

In Eq. (1), Δt, Δx, Δy and Δz give the space and time extent of the wave region, while $\Delta \nu$, Δk_x, Δk_y, and Δk_z give the values of the mean differences. Thus, the equation tells us that if the region where waves have not completely interfered one another out has an extent Δx along the x-axis, then we will have an average difference of at least $1/\Delta x$ in the x wave numbers of the waves in the system. We see that if the wave region is small, we shall require a large spread in wave-numbers, but if we are content with a widely spread out wave

[1] N. Bohr, *Nature, 121,* 580 (1928).

region, $1/\Delta x$ is small. In the limiting case of a wave region of indefinitely large extent, $\Delta k_x = (1/\Delta x) \to 0$; and therefore we need have no differences at all in x wave number. That is, we come back to the simple case of one wave, of indefinite extension, with no interference effects. Similar conclusions apply of course in the y and z dimensions. Eq. (1) exprpsses what is the most favorable condition for wave interference—with the phase constants of the waves adjusted for the smallest possible average spread in ν or k values for a space or time extension of the wave region.

Now let us suppose that we are considering electromagnetic or matter waves, to the frequency of which we can apply the equation $E=h\nu$. Since h is a constant, a spread or variation in ν, $\Delta\nu$, requires a spread in energy, ΔE; that is $\Delta\nu = 1/h(\Delta E)$. We substitute for $\Delta\nu$ in $\Delta\nu \cdot \Delta t = 1$, and find at once,

$$\Delta E \cdot \Delta t = h. \tag{2}$$

Eq. (2) tells us that if we have a group of energy-carrying waves, which we wish to confine within a time t, there is a limit to the "sharpness" of that time: specifically, Δt must at least be as great as $h/\Delta E$. If we wish to measure an energy at a given time—say we wish to measure the energy of an electron by having it move against an electrical field—we cannot measure that energy at an instant, but only over the time interval $\Delta t = h/\Delta E$. Clearly, the larger the energy involved, the smaller may be the Δt in which we measure the energy. If h were zero, as it effectively was taken to be in pre-quantum physics, we could determine energy at a time interval of arbitrary smallness, but in actuality the finite value of h sets a limit on the smallness of Δt. On the other hand, the value of h is sufficiently tiny from the point of view of every-day experience that we justifiably ignore the limitation on the time accuracy of an energy measurement. Even with an energy as small as one erg, the necessary Δt is only of the order of 10^{-27} seconds.

In Eq. (2) we have met an instance of the uncertainty principle, for we may properly interpret Δt and ΔE as "uncertainties" in our knowledge of time and energy. If we wish to determine an energy difference ΔE, then of necessity there is an uncertainty of at least $h/\Delta E$ in the time at which we can know that energy. That is, we cannot know that the particle or system involved has the energy change ΔE at the exact time t_0, but can only know this within a time uncertainty $\Delta t = h/\Delta E$. Or, likewise, if we wish to know the time of an event within an interval Δt, an energy $\Delta E = h/\Delta t$ must be involved in the time measurement.

We shall further consider the energy-time uncertainty relation in the next section. We can readily apply the $\Delta x \cdot \Delta k$ relations of

Eq. (1), and obtain an analogous uncertainty relation for momentum and position. We consider the equation $\Delta k_x \cdot \Delta x = 1$. By the de Broglie equation, wave-length λ_x and momentum p_x in the x-direction are in the relation $\lambda_x = h/p_x$. Since $k_x = 1/\lambda_x$, we may write $k_x = p_x/h$, or, because h is a constant, $\Delta k_x = \Delta p_x/h$. Substitution into $\Delta k_x \cdot \Delta x = 1$ then gives

$$\Delta p_x \cdot \Delta x = h. \tag{3}$$

There are similar relations in the y and z dimensions.

Equation (3) states that a determination of the position of a particle requires that there be a "spread" or uncertainty in the momentum of the particle. If the position determination is precise, so that Δx is small, then the uncertainty Δp_x in information about the momentum becomes large. Or, conversely, if a momentum determination is made on the particle, the particle waves and hence the particle are in a region $\Delta x = h/\Delta p_x$; if Δp_x is small, our uncertainty about the particle position becomes large.

The momentum-position uncertainty relation may be viewed as a direct consequence of the wave nature of matter. It is instructive, from this point of view, to consider the limiting cases of uncertainty. If the momentum of a particle is known precisely, the particle must have a single wave-length, in accordance with the de Broglie equation. There is then no possibility of interference among various waves of the particle, and the particle-wave extends indefinitely. Hence, the position uncertainty is indefinitely large. This result is as would be given by (3), for $\Delta p_x = 0$. However, if the position of the particle has been precisely determined at a point in space, then all wave-lengths are required for the interference that will cause the net wave amplitude to vanish everywhere except at the point of location of the particle. Since all wave-lengths are present all momenta are present, and we have complete uncertainty with respect to the momentum of the particle. Again, this result is what we gain from (3) when the space uncertainty Δx is zero.

The uncertainty relations of quantum mechanics may be approached by considering the effects of measurement on two quantities for which there are conjugate uncertainties such as energy and time, or location and momentum. Attempts to devise hypothetical experiments in which both quantities are simultaneously measured with an accuracy which is greater than is permitted by the principle have always failed.[2] This failure emphasizes the fact that the uncertainty relations arise not

[2]Many of these "*Gedanken* experiments" are described in the book by W. Heisenberg, *The Physical Principles of Quantum Theory*, translated by C. Eckhart and F. C. Hoyt, University of Chicago Press, 1930.

because of imperfections in our measuring techniques but are intrinsic in nature. Crudely speaking, we might say, for example, that because of its wave nature an electron cannot be localized at a single point if its momentum—which involves its wave nature—is also to be determined. There are physicists who believe that the uncertainty relations may be circumvented by future developments which will introduce new variables that permit precise determination of the conjugately uncertain quantities. It is indeed not easy, considering the surprises of the past, for one to predict the future of physical theory; but it is the case at present that the uncertainty relations are firmly fixed in our theory, and we may recall that the great changes in physics are usually developments in which the old becomes not wrong but a special and limited instance.

A rigorous derivation of the uncertainty relations and a more precise definition of the Δ's, within the framework of quantum theory, leads to, instead of (2) and (3), the following equations:

$$\Delta E \cdot \Delta t \geqq h/4\pi \tag{4}$$

$$\Delta p \cdot \Delta x \geqq h/4\pi. \tag{5}$$

In Eqs. (4) and (5), the presence of the inequality is equivalent to the statement that our Eq. (1) was for a most favorable circumstance. In other words, the product of uncertainties cannot be less than $h/4\pi$. The presence of the factor 4π in (4) and (5) need not concern us in our considerations of the uncertainty principle. It should also be pointed out that there are in quantum theory pairs of variables other than those of Eqs. (4) and (5) which are in an uncertainty relation to each other, but (4) and (5) will suffice for our needs.

We have written Δp in (5) without a sub-script. We shall consider Δp to be a linear momentum in any direction, taken to be the x direction in a suitably chosen coordinate system. If the velocity involved is such that the relativistic mass-velocity change is negligible, we may write Δp as $m \cdot \Delta v$, where v is velocity. Eq. (3) or (5) then tells us that we cannot precisely determine both the velocity and the location of a particle. In elementary discussions this result is often referred to as "the uncertainty principle," although more properly the principle is the fact that uncertainty relations like (4) and (5) do exist in physics.

It is interesting to note that the velocity-position uncertainty relation was presaged long ago by Zeno in his paradox of the arrow.[3] He argued that a moving body must always be at a precise set of points, and therefore could not go from set to set as required if it were moving. His paradoxical conclusion, then, was that the body cannot move.

[3]A discussion is given by R. Schlegel in "Quantum Mechanics and the Paradoxes of Zeno," *Amer. Scientist, 36*, 396 (1948).

There is a similarity to the conclusion that we gain from the uncertainty principle: if a particle is definitely located at a point, we can know nothing of its motion (the uncertainty in velocity is indefinitely large).

3. ENERGY AND TIME.

It is clear from what we have said about the uncertainty relation between energy and time that quantum theory introduces a limitation on the time concept. We discussed physical time in Chapter I as a variable which is quantitatively determined by uniform cyclic processes, and we gave no indication of the existence of any limitation on the accuracy of such determinations. And conceptually we tend to regard time as a continuous variable which may be "marked" or evaluated at as fine a point as desired, with respect to a given event. Quantum theory teaches us that the situation is not so simple as that. Our time measure must be in terms of physical processes, and the uncertainty relation prevents a physical process from being without time error, except in the limiting case where infinite energy is associated with the process.

A simple example is provided by a clock which makes use of electrical oscillations. In the ideal case we might have an oscillating circuit, the elements of which have been so physically refined that the circuit maintains a given frequency with complete uniformity, in terms of all criteria that can be applied. Receivers could tune in to this circuit and, by the use of suitable instruments, count and record the oscillations. We see, however, that even with this ideal clock we would have the quantum uncertainty. It would not be possible to record any fraction of an oscillation, for a single oscillation of the clock and of the receiver would be the energy unit for activation of the counter attached to the receiver. (If fractions of an oscillation were being counted, the original oscillation would be being divided into effective component oscillations, and the following argument would then apply to those oscillations.) For oscillations of frequency ν, the smallest discernible element of time is one period $\tau = 1/\nu$, or, the time uncertainty $\Delta t = 1/\nu$. However, an electromagnetic oscillation of frequency ν has an energy E/h associated with it. Hence, by substitution into $\Delta t = 1/\nu$ we find $\Delta t \cdot E = h$, in agreement with our results of the preceding section; the energy E associated with the oscillation is equivalent to the energy ΔE that must be available if the time uncertainty is to be as little as $h/\Delta E$.

A second physical example is provided by the situation of an energy transition in an atom. Suppose an atom is monitored by instruments which are set up to determine the frequency of any radiation emitted from it. We wish to know the time of a transition—resulting perhaps from the atomic nucleus going from a meta-stable to a stable state, or from an electron's falling from an excited to a ground state—and the transition will be signalled by the emission of radiation of frequency ν.

Again, however, the monitoring apparatus will not be able even approximately to determine the frequency unless it responds to at least one pulse of period $\tau=1/\nu$. There will then again be an uncertainty $\Delta t=\tau=1/\nu$ in the measurement of the time of the transition. It is true that the radiation receiver could respond to less than a full cycle of the emitted radiation: the receiver might be designed to give a signal on receiving, for instance, only a quarter cycle. But in this case there could not be a determination of the frequency of the radiation emitted by the atom. And if the frequency were not known, the energy of the transition would not be known.

Our second example illustrates then, that if the energy E of an event is to be determined, an uncertainty of h/E is involved for the time of the event. The first example illustrates that if we are to have a clock for the timing of events, we must have energy of at least E available if the time uncertainty is to have an ideal minimum value of h/E. We see from these examples that the quantum limitation on time is not a simple matter of saying that the accuracy of time measurement is limited to such and such a value. Rather, we have a limitation that is determined by any given physical situation. In principle, accuracy of time measurement could always be increased, because of the $1/E$ proportionality, by using processes of higher frequency. But the limitation is a significant one for physics on the atomic level, because any actual physical time measurement utilizes processes of limited frequency or energy.[4]

The relation of energy and time in quantum theory not only gives us the uncertainty of time measurement that we have discussed, but also leads us back to one of the primary generalizations that we can make about physical time.

Let us first note again that if there is no uncertainty in the energy measurement of an atomic process, the uncertainty in the time becomes indefinitely large. This result follows at once from Eq. (2) or (4) of the previous section. We can say, then, that if there is no uncertainty in energy measurement there is also no time measurement whatsoever. The case of simple electromagnetic radiation may again be used as an illustration. For radiation of a single frequency ν, there is some

[4]A discussion of quantum theory limitations on the accuracy of possible physical clocks is given in the paper, "Quantum Limitations of the Measurement of Space-Time Distances," by H. Salecker and E. P. Wigner, *Phys. Rev.*, *109*, 571 (1958).

energy uncertainty in the measurement, unless a wave train of infinite length is received. For there is always an uncertainty about one cycle; that is, the counting of cycles of the radiation, in a time t, always involves an uncertainty in the number of cycles, since an undetermined fraction of one cycle is not counted. Hence, if Δt is the (minimum) time of one cycle, which is $\tau = 1/\nu$, then $\Delta E = h/\Delta t = h\nu$. If n cycles are counted, the Δt for the event associated with the radiation is $n\tau = n/\nu$, and $\Delta E = h\nu/n$. In the limiting case of $\Delta E \rightarrow 0$, we must have $n \rightarrow \infty$, and therefore $\Delta t = n\tau \rightarrow \infty$, or, we have no measure at all of the time.

We may obtain this same result by using a formalism which will be well known to the reader who is acquainted with the mathematical symbolism of quantum theory. The *state* of a physical system can be described by a mathematical function known as the ψ function. At some time t_o, we may write $\psi_o = ue^{-iEt_o/\hbar}$, where u is a function of space coordinates only, $\hbar$ is $h/2\pi$, and E is the energy of the system. At some later time $(t+t_o)$,

$$\psi = \psi_o e^{-iEt/\hbar}, \text{ or} \tag{1}$$

$$\psi = ue^{-iE(t_o+t)/\hbar}. \tag{2}$$

We see from (1) or (2) that *if E is not changing with time,* the state function ψ at (t_o+t) differs from ψ at t_o only by a phase factor. And further, the product $\psi\bar{\psi}$, where $\bar{\psi}$ is the complex conjugate of ψ, will be the same at any value of t; and in quantum theory, $\psi\bar{\psi}$ gives the probability of finding the system in a given spatial region. In other words, the state of the system does not vary with the time if the energy is constant.[5]

If the energy of a system could change during the time t, or—what is equivalent—if there were an uncertainty in the energy, then we could not describe the state with functions, with constant E, of the form given by (1) and (2). The time-independence of the state when E is constant means that the system being described cannot in any way give a time measure; for if the ψ function gives a constant $\psi\bar{\psi}$ value, the observable properties of the system do not change. We have, then, by consideration of the properties of the function come to a result like the one that we gained by use of the uncertainty principle: namely, that a system of fixed energy has no properties for time measurement, or is, we might say, indifferent to time. The uncertainty-principle approach emphasizes our inability to make a time determination in an instance when the energy has been precisely determined,

[5]The material of this paragraph is presented in effect on p. 116 of *The Principles of Quantum Mechanics,* by P. A. M. Dirac, 3rd edition, Oxford University Press, 1947.

whereas the ψ-function approach emphasizes that the observable state of the system does not change when the energy is not changing.

The two statements which we have gained from the two approaches are in fact equivalent statements. From the uncertainty principle we gained the statement which I shall refer to as u, to wit: If the uncertainty about the energy of a system is zero, the time uncertainty is indefinitely large (or, the system provides no time measure). From the ψ function we gained the statement which I shall refer to as w: If the energy of a system is constant, there is no observable change of the function of the system in time (or, the system provides no time measure). It is a theorem[6] of logic that if there are two statements u and w, and the violation of u implies the violation of w, and also the violation of w implies the violation of u, then the statements are equivalent in content. We shall now show that our propositions u and w are equivalent by this criterion.

We violate u by assuming that although there is no uncertainty in energy, $\Delta E=0$, the time uncertainty in a measurement of the system has some finite value. Then, however, there must be a change in the observable state of the system, since otherwise there is no event of change in the system by which our interval of time can be marked. So, even if E is constant, $\psi\bar{\psi}$ cannot be constant. The statement w has therefore been violated in the violation of u. Next, we assume that although E is constant there is a change in $\psi\bar{\psi}$; we thus violate w. Then, even though the energy uncertainty $\Delta E=0$, there will not be an indefinitely large uncertainty in t, because there is an observable change of state which will provide some measure of t. Hence, u is violated if w is violated.

We have found then, for atomic phenomena as described by quantum theory, that a state of constant energy is one which provides no time measure. There is a noteworthy parallelism between this result and our conclusion about time, entropy, and change in Chapters I and II. We found there, considering large-scale, complex systems for which an entropy function may be determined, that if a system showed no entropy change it would be an atemporal system, considered in and of itself, without temporal quantity or direction. For an atomic process we find the same lack of temporal characteristics if the energy of the process is constant. The parallelism is quite exact if we take the entropy for a system which is at a uniform equilibrium temperature throughout—as would almost certainly be any system for which the entropy is not changing. Since entropy change $\Delta S=\Delta Q/T$, if $\Delta S=0$, then $\Delta Q=0$. But heat Q is a form of energy, so $\Delta Q=0$ is equivalent

[6]A discussion which does not presume a knowledge of symbolic logic is given by C. N. Hamtil, *Amer. Journ. Physics*, *22*, 93 (1954).

to E=constant. So, for both the atomic process and the uniform-temperature thermodynamic process the condition that energy is a constant gives a state which is atemporal, both in that it itself provides no time measure and in that it does not change with respect to any time measure that might be set up in its surrounding domain.

We further recall that we found that we were led in Chapter I by an analysis which was independent of any knowledge of physical theory, to the result that a world only of cyclical change would be lacking in temporality. Systems in which changes are uniformly cyclical—or, in the limiting case, in which there is no change at all—are physically characterized by being at constant energy. We have then a third line of evidence which suggests an association between a condition of constant energy in a system and lack of time characteristics for the system.

On the quantum level, the association is, we have seen, a rigorous one: constant energy of a state implies an atemporal state. On the level of large-scale, thermodynamic processes, constant energy appears to be a necessary condition for a process to be atemporal. The condition alone is not sufficient, because in a thermodynamic process many entities are involved, and their distribution at constant energy must also be such as to give a completely cyclic process, or a process with a zero entropy change. On the quantum level we are considering individual atomic processes, and the matter of arrangements of many entities among themselves does not come into consideration. The necessary and sufficient requirement of constant energy for an atemporal individual quantum process is, however, in agreement with the necessary requirement of constant energy that we found for large-scale atemporal processes. It is satisfying to find this indication of unity of nature between quantum-level and thermodynamic processes.

4. TIME AND DETERMINISM.

Before the day of the uncertainty principle it was generally accepted by physicists that the precise forms of any future states of a physical system were strictly determined by the present state of the system. A concomitant of this view was that the exact state of a system at some future time could be calculated or predicted—in principle at least—on the basis of knowledge of the laws of physics and of the present state of the system. The successes of physicists with the laws of mechanics seemed particularly to support this "deterministic" outlook, since mechanics in the eighteenth and nineteenth centuries came to be regarded as providing the keystones on which all of physical theory could be built.

We may express the physical determinism that preceded quantum theory in the following mechanical terms. Suppose a system contains i particles, each of mass m_i. And, let us say that at a time t_0 each particle has position coordinates x_i, y_i, z_i and velocity magnitudes $\dot{x}_i$, $\dot{y}_i$, $\dot{z}_i$. Finally, we assume that we know the nature of any physical forces that act between the particles, and that these forces are a function of the distances between any two particles, so that F_{ij} is the force acting between particles i and j. By application of the laws of Newtonian mechanics, with relativity theory corrections if necessary, we can now calculate the positions and velocities of the particles at any future time, and thereby predict the future state of the system. That is, we can actually make the calculation if the system is a very simple one, and in principle we presumably could calculate and predict for any system.

The deterministic view in physics is in effect a long extrapolation from the relatively simple situations that are calculable. The future then becomes a series of completely determined states, and the past similarly is a series of states any of which could be calculated, if desired, by application of the laws of physics to a given present state. The role of time, with this view of strict physical determinism, becomes curiously secondary in the bringing of new situations to the world. As time goes on, novel and unexpected situations appear, as we know from our experience, but if the total future is pre-determined by a given present state, these situations are in a sense always already existent in any given state. We say "the past is done," and in a sense we feel that past events are dull as compared with future ones because nothing can be changed about the past; its tragedies and errors are beyond possibility of correction. If determinism is strictly valid, the future is likewise set forever. Time may then be likened to the motion of a battery of floodlights which pass down a long series of sets; as each set is illuminated it is the present, but the others stand as future or past, altogether unaffected by the passage of the light.

Our experience and intuition lead us to a different view of the passage of time. The transition from present to future seems to us to have an element of creativity in it. The past is gone, but what the future shall be is in part determined by what we do in the present. The states of the present give birth to those of the future, and by acting in the present we can somewhat alter the future.

And yet, we know that the deterministic view is at least partially valid. We make calculations about the future in countless ways in every day life, on the basis of physical laws, and we find these predictions to be valid. Engineers, physicists, and astronomers make more refined predictions, which we likewise find to be trustworthy.

An involved electronic device generally performs as its manufacturer said that it would, and an eclipse of the sun begins on the second that was predicted by astronomers. Even complex biological organisms are found to behave according to rule in many ways. Everyone learns that in health a regimen of proper living will lead to a sense of well-being, and the physician learns the signs that all but certainly indicate the disease to follow.

There is, however, a difference between the existence of regularities in natural phenomena and the validity of strict determinism. The latter requires that every last atom of a system, as it were, is determined in its behavior forever. Even in pre-quantum physics, calculation soon reached a point with respect to any system beyond which further determination of a future state became impracticable.[7] The calculation of the effects of the various planets on each other in the solar system is immensely difficult, even for the strongest perturbing influences, and this in a system of just a few effective particle masses. One could not even begin to determine the exact path of an atom in a gas, where each atom is subjected to billions of different and varying forces in as short a period of time as a second.

The practical limitation on the exactness of calculation in physics gains the support of fundamental physical principle in the quantum theory; for the uncertainty principle prevents our gaining the information that is necessary for the calculation with unlimited accuracy of future states of a system. The extrapolation to unlimited determinism, from the ability to make good large-scale calculations of future behavior, is cut off then at a level which does leave some "indeterminacy" for atomic processes.

We readily see why the uncertainty principle limits calculations. We noted that strict determinism implies that by knowing the positions, velocities, and forces in a system of particles at time t_o we are able to calculate these quantities for any other time t. However, since $\Delta x_i \cdot m \cdot \Delta v_i = h$, we cannot simultaneously know position and velocity for a particle in any one coordinate dimension. Further, we therefore cannot know the distance between particles 'i' and 'j' which is required if we are to know F_{ij}. That is, we cannot exactly know the particle positions and separation distances at t_o if we are also to know the velocities. Hence we cannot gain the initial information with which to begin our calculation of the exact state of the system at a future time.[8] We may also express the lack of exact

[7]This point has been emphasized by Max Born; see his book, *The Natural Philosophy of Cause and Chance*, Oxford University Press, 1949.

[8]It is quite possible to make measurements from which one may infer both the position and momentum of a particle at some past instant. Thus, one may fix the position of an electron by noting the mark it makes in an interaction with a photographic plate, and the

initial information in terms of the energy-time uncertainty relation. If we knew the velocities and masses, for example, in a system of particles, we would then know the kinetic energies of the particles. Let us assume that no potential energies are involved, so that particle positions are not required for determination of the energy. Even so, if we did know the kinetic energies exactly we could not know the time t_o, and therefore would not precisely know the time of the state that we took to be the initial state of the system.

We emphasize again that the uncertainty principle is not a result of experimental inadequacy but is a result of the nature of things —essentially, of the wave-particle nature of matter and radiation and the non-zero value of Planck's constant h. There is, then, a physical principle that forbids our calculating future states to the accuracy, on the atomic level, that strict determinism requires. Physics has a working principle of determinism with respect to large-scale phenomena; this principle encourages the calculation of future and past states of systems, and it is justified by the uniformity of nature—the validity of physical law—that we do observe. But when we come to the level of individual atomic events we find that this principle is no longer correct, and nature does not present the information which would even in principle allow an exact determination of future behavior. We do have a "bridge" that enables us to pass between the atomic domain of indeterminacy and the large-scale world of approximate determinism: the statistical concepts of physics unify the two realms. We in principle cannot predict where a given diffracted electron will interact with a photographic plate, but we can predict the pattern of interactions that will be formed by an entire beam of electrons. Or, we do not know how a single molecule in a container of gas will move, but we do know that the average molecular behavior will give the relation between gas volume and pressure that is expressed by Boyle's Law.

The uncertainty principle has the very significant effect, then, of making untenable the strict determinism that appeared before quantum theory, as an obvious extension of the customary applications of physical law. Since many biological processes are apparently determined on the level of individual atomic events—the human eye, to give a simple example, can respond to as few as four or five quanta of visible radiation if they are incident within a sufficiently

---

momentum of the free electron after it leaves the plate could be determined with as high accuracy as desired. We would then be able to state with arbitrarily high accuracy the position and momentum of the electron at the instant that it left the plate. But such inferred knowledge of the past would not have any value in predicting future behavior; nor would we be able to extrapolate back to positions and momenta of the electron before it struck the plate, for the interaction with the plate would have involved a change in the momentum of the electron. We would not, in fact, have made a simultaneous measurement of the position and momentum of the electron.

short time interval—the quantum theory evidence against strict determinism has suggested new possibilities for the role of mind and "free will" in the determination of activities of man and other organisms.[9] And we can see that the existence of the uncertainty relations leads us to the problem of what are the factors which determine physical processes of all kinds (including biological ones). The strict-determinism view gave an answer to this problem, in that it implied that there was a predetermined mold into which physical processes must fit, exactly and without possibility of deviation. Now, we accept that there is a mold in an approximate and statistical manner, and we can still study the structure of that mold; but on the atomic level where the events that make up the larger processes are occurring we have lost the guidance of simple physical law. Yet, the larger processes show the regularities that we express in our equations, and further, there are coherent long-run processes of development among physical systems and biological organisms, and there are great cosmological patterns like the mutual expansion of the galaxies. Clearly, physical process is guided, but the developments of quantum theory suggest to us how little we do understand about that guidance; the simple view of strict physical determinism was only a partial and probably a highly superficial statement of how the course of physical process is determined.

In any event, physics no longer supports the view of time as merely the mode by which a predetermined order of things passes through the present. Rather, the physical processes which give rise to our time concept must be considered as being formed as they appear. Each present state, as it exists, determines the state that succeeds it, and time arises from the fact of change among these states. The successive new states follow from their immediate predecessors, in the approximate way that physics requires, but also we perceive that there is a level in the description of physical process or of physical states, beyond which our physical law cannot give exact prediction. Our theory has come to the recognition, then, of the same fact that we are led to by direct intuition: that nature is subtle and complex far beyond the possibility of being described by a physical science that is restricted within the domain of simple determinism.

5. RETRODICTION

The procedure of determining what the past states of a system have been, given a present state, is subject to the same modification

[9]Niels Bohr has particularly contributed to the discussion of the significance of the uncertainty principle, and has given an important general formulation known as the "complementarity principle." See his *Atomic Theory and the Description of Nature*, Cambridge University Press, 1934.

by the uncertainty relations as is the predicting of future states. And, we come to no new results concerning the role of determinism in physics by considering retrodiction. Nonetheless, between prediction and retrodiction there is an interesting difference which rests on a property of the irreversible changes of nature.[10]

We first note that for a time-reversible process the calculation of past and future states is the same. The system obeys an equation which is valid for either (+) or (−) values of t, as has been discussed in Chapter III, and there is no difference whatsoever in whether the calculation is going from the present to a future or a past state. In either case we do have the restriction, stated in the uncertainty relations, which prevents our gaining exact knowledge of a present state of a system; hence, there is the same lack of complete initial information in either case, and retrodiction can therefore be no more accurate than can prediction.

When we consider irreversible processes we find, however, that in general retrodiction is much less accurate than prediction. This fact is somewhat surprising, since we might intuitively expect that an estimate of the past of a system would be intrinsically more reliable than a prediction of the future, but one can readily find examples which show the more favored position of prediction. A metal bar, for instance, has been heated so that its temperature falls linearly from one end to the other. The bar is placed in surroundings of uniform temperature, and we are asked to predict its future temperature behavior. We have complete confidence in giving as a reply the heat distribution equation which states how the bar will approach a state of equal temperature throughout as time goes on. On the other hand, given a bar in a state of uniform temperature, we have no way of choosing between any of numerous possibilities for its temperature distribution at some previous time; the fact that its temperature is now uniform does not prohibit its previously having had some very sharp internal temperature differences which have disappeared as the bar came to temperature equilibrium. Or, as another example, we consider a mass of atoms which are in excited states; we can predict with assurance that if they are in "cool" surroundings, without the presence of exciting electrical fields, they will emit radiation and fall to their ground states, and will remain in those states. But given the atoms in their ground states at a present instant, we can only make a hopeful guess about what was the previous state of the atoms. They may have been in various configurations of excited states, or they may have been in their ground states since that earlier time for which we are retrodicting.

[10]Much of the material of this section has been suggested to me by a paper on prediction and retrodiction by Satosi Watanabe, *Rev. Mod. Physics, 27,* 179 (1955).

We may state the situation with respect to prediction and retrodiction as follows. A system is in a present state P, and we predict that it will go to some one of the future states F_i. Or, we guess that it came from an earlier state E_j. If the process of change of the state is a reversible one, we find that we can with equal assurance predict some one of the states F_i or retrodict one of the states E_j. If the process is irreversible we find that there are a large number of possible earlier states E_j, with no criteria for choosing among them. However, if the irreversible process is one for which we have a descriptive equation, we can predict with certainty some single future state F_i—neglecting both the possibility of statistical fluctuations in an irreversible process and the usual quantum uncertainties. We assume in this statement that we do not have information which enables us to give preference to one of the earlier states. That is, we assume that we must assign equal "weight" to each E_j. In practice it frequently is the case that we have independent information which enables us to select a given E_j with high probability. Thus, we may have been told that at the time t_e when the iron rod was put into uniform temperature surroundings it had for some previous time been in a steam bath at one end and in an ice bath at the other. We are then able to assign zero weight to all the possibilities for the state of the bar at t_e other than the one which would result from the steam-ice treatment.

But for the cases in which we are simply given the present state P of the system, with no information about past states, there is significant difference between the ability to make retrodictions and predictions. We do not require special knowledge about the weights of various future states in order to be able to make reasonable predictions, but we can only "blindly guess" if we do not have knowledge, beyond our information about P, which will give a basis for choosing among the various past states E_j. The effect of the environment must also of course be considered in this distinction. We are in effect taking the simple case in which the system may be considered to be in a uniform or non-disturbing environment, so that the prediction and retrodiction, over the time intervals involved, need not be concerned with special environmental effects.

The greater success in predicting than in retrodicting from a present state may be understood from a consideration of the properties of irreversible processes. We recall that for such processes there is always a tendency to move towards an equilibrium state, or, in terms of the H-theorem, there is a decrease to a minimum H-value as time goes on. We shall here neglect the possibilities of statistical fluctuations that violate the H-theorem and also we shall assume that the system is sufficiently isolated that there is no interference with the procession to equilibrium. The fact that the irreversible process tends to equilibrium gives a single end point for future behavior of the system, and once equilibrium has been reached there is no expectation of any further change. In contrast, however, when the system is in some state P (equilibrium or non-equilibrium) there are innumerable possibilities for the earlier states from which it came to P. A gas in a container may be put into any arbitrary state of non-uniform density, but from all of these states it will come to the equilibrium state of uniform density. Similarly, the iron bar may have been in one of countless different temperature states, from which it came to the equilibrium state P; or the group of atoms, in our example, could have come to their state P from any of a variety of excited states, but their future change will be toward their equilibrium ground state.

In statistical mechanical terms, the greater reliability of prediction is a consequence of the fact that most of the possible atomic configurations or "micro-states" of a given system are equilibrium states. Thus, suppose there are N possible micro-states in all for a system, and that m of these are equilibrium states. Also, in statistical mechanics each micro-state may be defined in a manner such that there is equal *a priori* probability for the system to go into each state. If m is a large majority of N, the equilibrium state will be much more probable than the state that corresponds to any other micro-state. If one must retrodict to a previous non-equilibrium state, there is only a $1/N$ chance that a given non-equilibrium state is the previous state E_j, but the probability is m/N (and this in practice is usually very close to 1) that the equilibrium state will be the future state.

Taking a related point of view, we can note that an equilibrium state is one of greater disorder (greater entropy) than a non-equilibrium state. We predict the state of disorder, but are not able to retrodict the particular state of higher order, for any of these states of higher order lead to the state of equilibrium. If all roads lead to Rome, we are able to predict that any traveller will end up in Rome. But we cannot with any high degree of probability look at the travellers in Rome and retrodict the road on which they came. Or, finally to put the matter still another way, a present state of a system is an actuality, and there are many possibilities from which that actuality may have come. The paths of development of the system converge however from the various possibilities to the equilibrium state. Since irreversible processes are "one-way" processes, prediction from an actual non-equilibrium present state is towards the single equilibrium state while retrodiction must look back to many possible previous states. In contrast, for the reversible processes the many-to-one relation does not obtain between the past and future states of a changing system, and as a result the difference between accuracy of prediction and retrodiction disappears.

6. TIME INVARIANCE

Until recently it has generally been assumed in quantum theory that individual physical processes on the atomic level are time-reversible. In a pioneer study, E. P. Wigner [11] considered the quantum-theoretical results of time reversal in atomic systems which were taken to be completely reversible in the usual sense of time-reversibility for physical procsses. In the theoretical studies of interactions between elementary particles, physicists have frequently made use of the hypothesis that the reverse of any reaction is as probable as the reaction. Thus, if a hyperon decays into a proton and a pi-meson, then a properly energetic pi-meson plus a proton collision should result in formation of a hyperon (with a degree of probability that is calculated from the life-time of the hyperon).

The property of time reversibility for the processes of nature on the atomic and elementary particle level is usually referred to as "time invariance." The term indicates that the processes are unchanged in their nature and probability of occurrence by a change in time direction; that is, they are invariant under time reversal, or they would be indifferent to whether they are going "forward" with increasing t or "backward" with decreasing t.

In 1956 there began to be concern with whether or not time invariance is a valid assumption, at least for all kinds of quantum-level processes. This questioning arose because in that year another well-accepted kind of physical invariance was shown not to be valid for all processes. It was found that some elementary physical processes have a fundamental spatial asymmetry, such that a process which occurs in a given experimental arrangement will not occur in an arrangement which is a mirror image of the first; or, a process that can be described in a right-handed coordinate system will not occur in a way that allows it to be similarly described in a left-handed system.[12] Thus, it was found that when cobalt-60 nuclei are vertically aligned by application of a magnetic field, electrons are emitted from the nuclei essentially in only the up or the down direction.

The nucleus of cobalt-60 is unstable, and decays into nickel-60 as a result of emitting an electron and a neutrino. The cobalt nucleus

[11] E. P. Wigner, *Göttinger Nachr.* (*Math. Naturw. Klasse*), 1932, p. 546.

[12] Two popular review discussions are "Fall of Parity: Recent Discoveries Related to Symmetry of Laws of Nature," by L S. Rodberg and V. F. Weisskopf, *Science, 125,* 627 (1957) and "The Overthrow of Partiy" by Philip Morrison, *Scientific American,* April, 1957.

has a spin and associated magnetic field, and thereby can be aligned along the axis of its spin in an external magnetic field. If the nuclei are thought of as being in a plane, perpendicular to the magnetic field direction, that plane is a symmetry plane as far as the nuclei are concerned. The applied magnetic field may be considered to result from electric current in a flat coil which surrounds the plane, and there is then a spatial asymmetry associated with the current and the nuclear spins; thus, from one side of the plane the current would appear to have a clock-wise direction while from the other it would appear to move counter clock-wise. This element of asymmetry apparently does mean something to the decay processes within the cobalt nucleus, or to the emitted electrons and neutrinos, since the electrons are observed, subject to expected experimental variations, to come out only from one side of the plane of atoms. The observed asymmetry may be a result of an inherent spin-direction relation in the emitted neutrinos, such that the neutrino with, say, a right-handed spin can move only in the forward direction of a right-handed screw, while the anti-neutrino has a left-handed spin and moves forward only with the rotation sense of the forward motion of a left-handed screw. The neutrino emitted from the cobalt-60 could be expected to have the same spin sense as the nucleus, and it would then actually be an anti-(or left-handed) neutrino. These neutrinos would come out only on one side of the plane of the nuclei, and electrons can be assumed mostly to come in the same direction as the neutrinos. The observed asymmetry is thus explained. The emission "polarization" of a plane of nuclei has also been observed for cobalt-58 nuclei. These nuclei decay by positron and (right-handed) neutrino emission, and the direction of positron emission is found to be just the opposite of that for cobalt-60.

An asymmetry has also been observed for the decay of a mu-meson into an electron and two neutrinos. The electrons are found to be emitted in a preferred direction from mu-mesons that are virtually at rest. Again, the mu-meson does have a spin, and presumably that spin, via the mechanism of the accompanying neutrinos or in some other fashion, determines the direction of electron motion.

Whatever the confirmed explanation may ultimately come to be, it is clear that physicists have found that some natural processes involving electrons are not indifferent to a "reflection" of spatial coordinates, as once was assumed. It is natural to ask, Is there possibly some asymmetry also with respect to the time coordinate of elementary physical processes? There is a considerable body of experimental evidence which suggests that there may be at least slight temporal irreversibility in the high-energy interactions among the elementary particles of nature. The interpretation of the experiments is, however, strongly dependent on theoretical assumptions which are themselves not beyond question, and it can hardly be said that temporal asymmetry has been established. The problem is currently (1968) one of high interest among physicists.

The finding of deviations from time invariance on the level of elementary physical processes would constitute a significant change in our concept of physical time. We have seen that time invariance clearly is not found everywhere in nature: the irreversible processes of many-element or thermodynamic phenomena are not invariant under hypothetical time reversal. On the other hand, the elementary processes which constitute these irreversible processes are generally taken to be time-reversible, or, we could say, time invariant. The progressive, irreversibly changing aspects arise, then, because of the operation of processes that are described by the *H*-theorem; reversible processes can combine to give progressive changes that give rise to a sense of time direction. But, if on the level of the elementary processes, hitherto taken to be time-reversible, we should find a lack of such reversibility, then we would have an on-going temporal process at the fundamental level of natural process. Our concept of time's moving forward could then perhaps come to have a further basis in elementary physical processes.

At the far other end of the scale from the elementary particle processes we do in fact find a process which seems to have a single time direction and yet is perhaps not of the class of irreversible thermodynamic processes. I refer to the motions of mutual recession among the galaxies. But as yet we have virtually no linkage in physical theory between the very large of cosmological phenomena and either the middle ground of thermodynamic processes or the very small of elementary particle processes.

Chapter X

A Dialogue on Time

Peter, a philosopher, Roberto, a physicist, and Edwin, a physical chemist, have seated themselves on the porch of Peter's summer home. Pine and cedar trees come close to the cottage, and through a break in them one can see a stretch of beach sand and the blue waves of Lake Michigan. Peter is half reclining on a divan, Edwin is in a cushioned cane chair, and Roberto is enjoying an easy motion in a rocking chair.

Edwin says to Roberto, "You have been working on the problem of time in physics. This sounds intriguing indeed—are you learning anything new?"

"Well, yes, I feel that at least my own education about the questions concerning time has been much improved," Roberto replied. As he paused to form his next statement Peter broke in:

"We philosophers know that Newton gave physicists a time variable that was clear and absolute, a uniform variable to which events could be referred and yet that in its nature was outside physical happenings. Then Einstein showed that time is affected by motion, and I understand that physicists now accept that time arises *from* events, rather than that time exists before events. Have there been any further developments in the physicists' notion of time?"

"Certainly it is the case," Roberto replied, "that Einstein and his special relativity theory made the big change in physicists' concept of time. For centuries there had been no need, at least as far as physics was concerned, to question the idea that a universal and uniform time measure exists. In suggesting that relative motion can affect time rate, Einstein very sharply brought physicists to the idea that time is not independent of physical processes. I don't know of anything in physics that suggests any new change from that basic point of view, although as our physical knowledge grows we continue to learn new things about the time characteristics of the natural world.

"In my own thinking about time," he continued, "I have been

trying to see what are the characteristics of physical processes that cause us to have a concept of time at all."

"You mean," asked Edwin, "time's going along with the entropy increase of natural processes?"

"Yes, I do think that entropy increase is associated with the forward direction of time," Roberto replied. "But even before we make the measurements that enable us to calculate entropy changes we know that time is in the world. The very fact of change in nature is at the root, I believe, of our having a time concept. The universe is in one state, and then again in a different state, and we distinguish the two states as being at different times. It so happens that besides presenting a world of successively different states, nature also contains processes of uniform change, especially cyclic processes, such as planets moving in fixed orbits, or pendula regularly swinging back and forth, or atoms oscillating in molecules. These processes serve as bases for clocks, by which we can define time units and quantify the time between different states."

"How do you then handle the earlier-later relation?" asked Peter. "Do you get away from an ultimate appeal to man's psychological time sense?"

"I don't think," said Roberto, "that the problem of time's direction is so difficult. We do in fact of course gain our sense of past and future from our own experience; the existence of memory and the on-going processes of our own bodies must be important elements in our having that sense. But physically we can define time order among the states of a system if we have contact with the system. Past states leave a physical record, observable in the present, but future states do not; as the present moves on, the group of past states increases, and the time of any given past state may be found by using a clock which tells us the number of time units between the present and the event signifying a given past state. Thus, using a clock you can tell that the rain of yesterday was not so far in the past as the snow of five months ago."

"But," objected Peter, "what if you are not in contact, as you put it, with the system?"

"Then to be sure, you must use some criterion internal to the system to determine the past-future relation between two inferred states. The generalization that entropy of an entire system increases could then be very useful. But direct analogy with past experience might also suffice. If you see the remnants of a garden corn patch, you feel confident that states of the patch in which the corn was in the form of green shoots preceded in time a set of states in which the corn was tall and brown."

"You are saying then," Peter enunciated thoughtfully, "that the changing states of the world, and not our psychological sense of earlier and later, provide the basis of time."

"That is right," said Roberto, "with also the addition of the uniform cyclic processes that enable us to make clocks—which give a time measure of both the uniform and non-uniform aspects of physical change."

Peter quickened and exclaimed, "Ah, yes, but what then of the case where there is no change to ever new states; could there be a world of cyclic processes only?"

Roberto replied that our world was at least not obviously such a world, but that he knew of no physical argument against the universe having this characteristic.

"I have read anthropologists' accounts," Peter said, "of societies which are virtually static socially. The seasons and the generations of man repeat more or less uniformly, and often men in these societies have very little time-sense; certainly not the keen sense of history and of past and future that we of Western culture have."

At this point Peter and Roberto turned to Edwin, who had leaned forward in his chair to speak. "Your statement, Roberto, of why we have a time concept is perhaps all right," he said, "and yet I can't think that it strikes very deeply into what time is. You speak of the universe's going from state to state, and of these states being different. There is just where the point—and mystery—of time lies. The universe changes, and time is a process that brings new out of old. But there is continuity too in the process: the old is inextricably bound up in the present, and both past and present go into the future. Time gives the past its property of being complete and untouchable, and the future its property of being unknown and incomplete, and yet too, time binds past and future into a coherent scheme of things. Your statement is too simple. Time is not something that just occurs to us because the world is changing, but rather time is in the nature of the changing order of things. Further, the universe is not a simple progression of states, with the present sharply set off from past states. The universe is a chemical and physical process, and the effect of a given state only slowly recedes; the division between past and present is arbitrary, and to make the division actually misses much of the significance of time."

"Bravo," said Peter. "You give a good criticism, Ed, of what Whitehead would have called a simple-minded materialism in Roberto."

"Very likely I am simple-minded," Roberto said mildly, "but in reply to Edwin I want to say that in proposing progressive and cyclic change as the elements of nature that give rise to time I don't by any means think that I have exhausted the subject. Of course, all of the problems associated with natural change do remain. Chemists, biochemists, biologists, geologists, astronomers—they are all trying to understand various processes of change and development. We in physics, in our attempts to elucidate the elementary particles and forces, and what results from combinations of them, are trying to supply fundamental information for the understanding of the processes of nature. To state, as I have, those general characteristics of natural changes which give rise to time is just (he smiled at Peter) to make a philosopher's generalization. The infinite details of physical, and chemical, and biological process remain for patient investigation.

"I suppose," Roberto continued, "that I have been guilty of oversimplification in talking about states—past, present, and future—of a system, or of the entire universe. Actually, quantum physics has taught us that we cannot specify a state precisely at a given time; if the time is exact, we cannot state the energy of the state, and in no case can we give precise, simultaneous specification of the positions and velocities of the elements of the system. So, a 'state of a system' is a somewhat vague concept, at least on an atomic level, and in many cases it can only be given in statistical terms. Again, if we do obtain a precise determination of the energy of a system, we cannot know exactly the time at which the system has that energy. Actually, then, the nature of the physical world does itself require that we cannot set a time as the present and mark off the past or future states of the world from that present state. The whole question of determinism enters here, but I don't think we have to get into that." He paused, and then concluded, "In a word, when we come to the atomic or sub-atomic levels, where natural processes are at root occurring, we do find that we cannot sharply distinguish among past, present, and future states. And yet, these distinctions are valid among every-day or laboratory events as long as one does not wish to make time measurements of very high accuracy. Such measurements are not involved in the experience which gives us time a concept, and therefore I think one is justified in speaking of the past, present, and future of a given system."

Edwin spoke again. "I can understand what you are saying, and yet I am not altogether satisfied. There is the matter of relativity theory: how can you so easily speak, for instance, of the present state of the universe? The Lorentz equations tell us at once that two events, some distance apart, have different time relations for different observers. Say the events are simultaneous for *A*. Then for *B*, who is moving with respect to *A*, one of the events is earlier than the other. So, what can 'present state' of a system mean—is it different for each observer?"

"Yes," replied Roberto, "in part at least that must be the case, and yet I don't see any problem there. As you know, a result of relativity theory is that we try to describe physical phenomena in equations which are of the same form for all observers, even though relatively moving observers will find different numerical values when they specify their own velocity conditions in the equations."

Peter broke in, "But this seems to mean that you have not one series of states for the universe but an infinite number of different series, that is, a different one for each of the infinite number of possible observers."

"Yes and no," said Roberto. "Suppose we think of our system as being an electron and a line of space points on one of which the electron is situated at rest, for observer *A*. Then events at separated space points that are simultaneous to *A* are at different moments for the moving observer *B*. We know that application of this relativity theory consequence to the energy-frequency of the electron results in the electron's presenting a wave form—a de Broglie wave—to the observer *B* if he sets up interference apparatus for observation of the electron. So in this case our system does present different series of states to different observers. And it is likewise with all the phenomena that involve high-velocity physical interactions: the time characteristics of a given physical process vary in different physical interactions, depending on the relative velocity of the system undergoing the process."

"But," continued Roberto, "when we consider systems which are not in any kind of physical interaction with an observer, I am not so sure that it is legitimate to apply the equations for relativistic time effects. I think of two hypothetical galaxies which—by rare chance let us say—are completely identical in physical content and structure, as judged by an observer when at rest with respect to each galaxy. The only physical interaction between the galaxies is, we shall say, the trivial one of exchange of communications by radio-waves between observers in the galaxies. The ascription of relativistic time effects to the galaxies would require that all of the processes in one run at a slower rate than in the other, assuming, that is, that there is a relative speed between the galaxies. If the galaxies passed each other, and the radio-communicators told each other that the galaxies were at a similar stage of development with respect to some process, then as the galaxies went away from each other, say at a speed one-half that of light, the observers would, if the application of relativity is valid, be reporting different rates of development for that process. Well, you ask, as do I, in which of the two galaxies would the rate be slower? Those who accept that there would be a difference are forced here to rely on some kind of preferred coordinate system. The other galaxies of the universe, for example, may in effect determine a master-system with respect to which one of the galaxies in some physically meaningful way has a smaller velocity than the other."

Roberto paused, and then continued. "This may be the way it is, and yet, I like another alternative—namely, that relativity theory applies to physically interacting systems, and that when two systems are physically independent the time rates of their internal processes are not affected by relative motion between the systems. On this view, our two similar galaxies would have the same rates of development, and in general all observers would find by their own clocks a single time rate for a given process that is not in essential interaction with them. Then the universe, or a system, would present a common present state for all observers." He stopped, and then said in conclusion, "Again, of course, with the understanding that we exclude situations of high velocity physical interaction; the processes in those situations do display relativistic time changes."

"I gather," said Edwin, "that you are now making the distinction that we talked about some time ago, between what you called Clausius processes and Lorentz processes. Developmental changes within a galactic system you would say, I suppose, are Clausius processes that are unaffected by relative motion of the system with respect to another."

"Yes, that would be my idea," Roberto replied.

"And how," asked Peter, "do you handle the matter of the velocity of light? I have been told that the whole point of relativity theory is just that clocks and other processes of two systems must have different time rates, just so that observers on the two systems would find the same speed c for light from a given source."

"Ah, in a way the speed of light is the crux of the whole matter," Roberto quickly answered. "But I suggest that it is the photons of light that undergo the Lorentz transformations—that is, the relativistic time and space alterations—rather than all the clocks and meter sticks on two relatively moving systems. Somehow this seems far simpler and more likely: photons adjust their frequencies and wave-lengths, when they interact with observers, so as to give their speed between the source and the observer the value c. Photons are then of the class of Lorentz processes, which are observed in direct physical interaction and show a change in temporal properties between relatively moving observers. For photons the frequency and wave-length changes are directly observed in the relativistic Doppler effect: light does show relativistic frequency changes if observed from a source at rest and then from similar sources moving at a high speed."

"Do you mean then," asked Peter, "that only certain kinds of processes could serve as good clocks?"

"The processes which can be observed from all coordinate systems—without interaction with the processes—would be proper clocks on my suggestion. The clocks that we use would generally fall into this category of processes. The internal mechanism of a spring clock, for example, would not be affected by uniform motion, and observations of the clock from any coordinate system would be made, say, by light rays reflected from the clock face. The light ray properties would certainly be responsive to relativistic velocities between clock and observer, but I would expect similar spring clocks in all coordinate systems to maintain the same rate."

"But the electromagnetic oscillations of light rays would not themselves be a good clock process?" Peter suggested.

"No, indeed not; observers with differing velocities relative to a light source would find different frequencies for the light they observed. But their various Clausius process clocks would, I think, keep a common time, by reference to which they would observe the varying Doppler shifts that relativity theory requires for light coming from the source to them."

"What you are suggesting then," said Edwin, "is that in a group of coordinate systems moving with various speeds with respect to the source, physically similar clocks or other processes that are internal to the system all have the same time rate. Light from a single source has relativistic variation from system to system, because of transformations in the light rather than in the clocks. On the usual view, there are genuine differences in rate of clock process—even for physically similar clocks—as we go from system to system, and it is for that reason that the light from one source has relativistic frequency differences for different observers."

As Edwin finished speaking, and as Roberto nodded in agreement, the discussion was interrupted by the wives of the three men and their children coming onto the porch. The children wanted an afternoon swim, and the mothers were going to the beach with them. Did the men want to come too? There was some talk about the coolness of the air, and about the warmth of the water. Finally the men decided that they would not swim today, and then they relaxed for further talk.

"The role of cyclical and non-cyclical processes in giving us time intrigues me," said Peter. "I can see that much of the interest and variety of life arises because there are both kinds: progressive non-cyclical changes keep the future interesting and challenging, while cyclical processes bring us the familiar and expected."

"Yes," said Roberto, "it is good always to know that Christmas will come again, and yet also that five years from now our lives may be different, and we hope better, than they are today."

The three sat silent for a moment, and then Edwin proposed, "I should think that we could say that the cyclical processes supply stability to our world—as well as giving us quantitative time measures as Roberto has told us. The solar system is stable with the regular motions of its members. The stable chemical compounds are made up of entities that have periodic, prescribed motions with respect to each other. I think of the permanence of a rock, or for that matter of an animal organism. Everywhere in the universe there seems to be motion and activity, and yet, because there are uniform, cyclic processes our natural world is quasi-permanent and stable."

"Your point is a good one," Roberto said. "It is interesting to think of what the world would be like without the existence of both cyclical and non-cyclical processes. Peter has pointed out the timeless-ness of a world consisting only of well defined cyclical processes. If there were non-cyclical changes only, I suppose nature would have a chaotic quality, or at least would present a formless *chiaroscuro* of ever becoming something different. Yet, we cannot rule out the alternative of a natural world of uniform progressive change—no cycles, but still everywhere a steady drift in some consistent manner or direction."

"You mean," said Peter with a laugh, "a world that is in all ways getting a little worse each second—or a little better."

"I guess that's the idea," Roberto said. "Actually, a world in which all processes were like the apparently uniform expansion of the galactic universe is what I had in mind."

Edwin seemed to consider this possibility, but then said, "To come back to the world as we do find it, it appears rather striking that we have just the conditions that permit development and evolution, in all realms of nature. The cyclic processes give us enduring systems, but because of the possibilities and tendencies toward non-cyclic change, there can be modification of these systems. The stars and galaxies are permanent indeed by our human standards, yet even they undergo progressive development. Among living things, we all well know how it is: each human being is about as were his parents, so that the cycle of birth and death maintains the human species, and yet, there is the possibility of change that allows progressive evolution of the species. Nature seems to have ingeniously combined her processes, so as to avoid the chaos that would come with no continuity of system and also to have the non-cyclic change that allows for development—or evolution—of systems."

Peter added: "It is striking indeed. Using the terms of aesthetics one might say that nature presents both unity and diversity: non-cyclical change insures the introduction of new elements, of novelty, and the cyclical processes provide the stable patterns that give coherence and unity. Thus a man in his life can have a diversity of experience which is bound into the continuity of his being and personality."

"I wonder," Roberto said, "if physics does not suggest some further insight into this situation. The distinction between cyclic and non-cyclical processes seems to be important for understanding what are the properties of nature that cause us to have our time concept, but in physics we have come to be more concerned with the distinction between time-reversible and time-irreversible processes. The reversible ones are those which, roughly speaking, can go either way as far as the laws of physics are concerned. A particle which moves reversibly will exactly retrace its path if its velocity is reversed; or, the equation describing the motion gives the same path excepting for a direction reversal, whether it is solved for time going forward or time going backward. The motions of the elementary particles of physics are largely described by those time-reversible equations, and also even massive isolated bodies like the planets are time-reversible in their motions—at least to a very high degree of accuracy. The time-irreversible processes seem to be those which involve interactions among many particles, and they are processes which go only one way: processes of diffusion, mixing, destruction, and the like. If a bottle of milk is dropped and broken, the process is irreversible, for it is not in keeping with the laws of physics that the broken parts of the bottle will again coalesce and the milk draw itself into the bottle. However, it was a remarkable achievement of the physicist Boltzmann that he showed that time-irreversible behavior of a system will result from the motions of many particles of the system, even though the processes of the individual particles are time-reversible."

Edwin commented that also in thermodynamics the terms "reversible process" and "irreversible process" are used, but with reference to the entropy change involved in the process rather than to the time properties.

"Yes," Roberto went on, "the many-particle time-irreversible process would always be a thermodynamic process, since it does involve a macro-system of many particles, and it would be irreversible too in the sense of proceeding with some net entropy increase. The reversible thermodynamic process however would be different from the time-reversible process, which is generally a process involving one or only a few separated entities, and would not be spoken of at all as a thermodynamic system. But actually, thermodynamic reversibility is a limiting or idealized case, is it not?"

"Quite right," Edwin answered. "We do not know of any natural thermodynamic processes that are clearly reversible in that they go on with no net entropy increase. Although some people have spoken of the expansion of the universe as being a constant-entropy and hence thermodynamically reversible process."

"That one is a puzzler," Roberto added. "It has also been spoken of as being time-irreversible, and as we judge the expansion process now to be it would certainly not seem to be of the class of time-reversible processes."

Peter broke in to say to Roberto, "Do you mean that all of the motions of the elementary particles are time-reversible? Is there then a fundamental stratum of these time-indifferent or time-less processes out of which our world of time and change is formed?"

"That is not a question that I can answer with any certainty. As I said, it has seemed a fair assumption so far, that the elementary particle interactions are time-reversible in a general way. But, it is the case that the fundamental equations of physics do not require such reversibility; forces or fields which give time-irreversibility on a fundamental level might be discovered. And, after all, we do have galactic expansion on the cosmological level, and as we were saying that is not a time-reversible process."

"Where," asked Peter, "do our cyclic and non-cyclic processes come in?"

"Now don't be impatient," Roberto said, with a friendly grin. "I am about to come to them again."

He continued, "Certainly the cyclic processes that serve as bases for ordinary clocks are thermodynamically irreversible processes that proceed with a net entropy increase, resulting from the combination of the entropy changes in both the clock process and its surroundings. But the cyclic processes which Edwin referred to as giving permanence and stability to nature—the motions of the electrons and nucleons in atoms and molecules and the closed orbits of the solar system—these are time-reversible processes. And so, Peter, I suppose that you could say that the time-indifferent processes, as you called them, are a basis for existence of stability in nature. But again, atoms moving in time-reversible paths do combine into many-atom systems that behave irreversibly in time and irreversibly thermodynamically. Thus do the non-cyclical processes arise: the processes of progressive change—of increasing disorder and increasing entropy in thermodynamic systems—that give nature the non-repeating patterns of change. And these patterns give rise to our sense of a forward moving and never repeating time in the universe."

"But also," Edwin broke in to add, "we must not forget that the many-atom systems also give rise to irreversible processes that are cyclical. You mentioned the ordinary cyclical clock, which does proceed irreversibly, with net entropy increase. Nature herself establishes stable thermodynamic systems, and in doing so establishes increased order locally. Atoms fall into the patterns of crystal lattices. And, complex biological organisms are formed. In them there are both the approximately cyclical behavior of birth, life, and death, generation after generation, and also the remarkable property of development of increased complexity among species of organisms. Or, for that matter, it is remarkable that any one individual organism grows, thus bringing about a high degree of order as tissue, nerve, and organ are fabricated from food stuffs of simpler organization."

Edwin then added, somewhat as an afterthought, "But of course we must not think that there is any violation of the general principle of entropy increase, or disorder increase; the individual organism or system that gains in order does so at the expense of a greater increase in disorder in the surroundings."

Roberto held out his hands as if to enclose a volume and said, "We can sum up then by saying that the time-reversible processes include some inherently cyclic processes that are apparently essential elements in the permanency that we find in nature. And," he added parenthetically, "we are learning how to use these basically time-reversible processes as time-measuring processes, in our atomic clocks." He continued, "The time-reversible behavior of atomic particles in systems of many particles is generally commuted into behavior of the system that is irreversible, both time-wise and thermodynamically. And yet, the natural laws are such that among these irreversible processes there may be thermodynamic systems that are cyclic; and also there arise systems that by interacting with other systems are able progressively to change to states of lower entropy and higher order. All this seems fairly tidy, but we are still left with the possibility of fundamental time-irreversible processes; the putative existence of the expansion of the universe makes one think that on the basic physical level there may be something more than simple time-reversibility."

After a pause Peter commented, "In any event I am impressed by the fact that there are the cyclic processes that are out of the stream of non-cyclic change in the universe. Poets and novelists have long written about time-less aspects of nature; it is good to realize that physicists too find aspects of the world which do not share the onward motion of clock and calendar time."

"All we are saying," Roberto added, "is that there are in nature systems whose processes of change are always bringing them to new states, other systems whose processes of change are cyclical and repetitive, and some systems that can I suppose be said to change virtually not at all."

"True," said Peter, "but the progressively changing processes, and the time concept, are so important to us, that there is a tendency to think that time, as it were, is moving throughout every process of nature."

At this point the conversation was interrupted by the pawing of Peter's dog on the outside of the screen door of the porch. Peter let him in, and the dog made a bound toward Roberto, and then to Edwin. After receiving a little attention from each of those two gentlemen, he settled down close to Peter, at a spot where his master could give him an occasional pat on the head or tug on the ear.

Peter then continued on the same topic that he had been discussing. "I even find it very unsatisfactory," he said, "to think of any part of the universe as progressively changing into ever new states. It scarcely seems that there can be such an openness of possibility in the universe. My judgment is perhaps entirely emotional in this respect—or at best no more than metaphysical—but I feel that there should be a natural limit to any process of development. After all, excepting perhaps in the matter of spatial extension, the universe that we experience is definite and limited. I know from chemistry that there are only a certain number of ways in which a given group of atoms can be combined, or, to give another example, a tree can grow and grow, but eventually there is a limit to its size, and in time it does deteriorate and fall back to the earth from which it came. The vast universe of stars and galaxies of stars is of course incomprehensibly larger, but still it is all one and the same natural universe, and so I would think that there are limits inherent in the universe."

Roberto was clearly interested in what Peter had said, and yet spoke in a cautious manner: "What you have expressed may be an idea of great significance." He then went on, "Perhaps it would be worthwhile to look again at the process of expansion of the universe, for there we seem to be faced with the problem of a progressive kind of change extending indefinitely into the future. I think first we should realize that mutual recession of the galaxies does not necessarily imply any effect within our own galaxy. It may be that the processes of formation, development, and eventual dissipation of stars would continue indefinitely within our galaxy, even though the expansion of the universe leaves it virtually a lone island universe. After all, it does contain billions and billions of stars, and is some hundred thousand light years across."

"You mean," said Peter, "that our galaxy provides an ample

stage for us, and we should perhaps be content to restrict our universe to it."

"Yes. My notion was that the galaxy is a physically isolated unit, and that our extrapolations to what will in any way physically affect it could be confined to the galaxy. But yet, I don't believe that. Astronomers tell us that the galaxies themselves appear to be in various stages of development; so it hardly seems likely that our own galaxy is going to continue to be more or less as we today find it to be. And, the coherence of the mutual expansion among the galaxies certainly argues for there being a physical relation between them: in some way they are part of a single physical entity."

Roberto continued, "Well, what can we say then about the expansion? It clearly does seem to exist—it is remarkable how radiation consistently gives the Doppler red shift. I do remember once hearing an expert in cosmology say that it might just be that the red shift did not actually arise from recessional motion, and that this possibility constituted an occupational hazard for cosmologists. But physically we know of absolutely nothing else that would give the shift, and I think we must take it that the recessional galactic motions are real.

"And yet," he continued, "we do not for certain know anything more about the motions than that they have the regular distance-velocity relation that we do observe. If the galaxies are assumed to have had their currently observed velocities throughout their period of motion, then we can extrapolate backwards in time and find that the galaxies must all have had an origin in the same spatial region, astronomically speaking, and at about the same epoch of time. Of course, the assumption of a constant velocity of recession for each galaxy is without any particular justification. It would be equally reasonable, perhaps, to assume that the universal expansion that we now observe is changing in rate, with an acceleration that began with a zero expansion velocity at some time past. Or, for that matter, the acceleration may be decreasing from previously higher rates. The cosmological models which we can construct on the basis of physical theory do allow for a variety of kinds of expansion behavior, depending on the kinds of physical parameters that are assumed for the model. We are, however, at the point in astronomy where observation is just beginning to give data that can be expected to allow for decision for or against at least some of the models."

Peter broke in to ask, "But in any case, we must accept a world in which the galaxies are racing away from each other, giving us more and more empty space?"

"That is right," Roberto said, "excepting, to be sure, that there is the very interesting possibility provided by the steady-state theory.

According to that hypothesis, as you probably know, there is compensation for the expansion of the galaxies: new matter appears in space, so that the total amount of galactic matter remains constant in any large astronomical volume."

"Then," said Peter, "it seems to me that infinite progressions are compounded. The process of galactic recession goes on infinitely, and also matter is formed without end in an infinite universe. We lose all sense of comprehensibility of the universe."

"Yes, so it would be," Roberto replied. "But remember that each observable universe would remain approximately uniform in its galactic population, and there would not be progressive change in the large-scale features of that universe. In a sense you would escape the process of progressive change, and would live in an observable universe of galaxies that could perhaps be called time-less. But let's return to the orthodox cosmological models based on the observed recessional motions. Our concept then is of an expansion process that is increasing the distance between the galaxies and therefore giving us a progressively larger galactic universe. There is perhaps little ground here, Peter, for your suggestion of finitude for the natural world—although I believe you did say that you excluded spatial extension from your hypothesis. Still—"

Roberto hesitated for a few seconds and then went on to say, "Even with respect to spatial extension there are some grounds for rejecting a universe of infinite extent. It is impressive, you know, that in looking at the skies at night one sees a dark background for the stars. It was shown more than a hundred years ago by Olbers that if the stars extended indefinitely in space, the sky would be ablaze with light, with an intensity somewhat as if our sun were at every spot in the sky. The argument is, as the volume of a space increases, the number of enclosed stars increases, but whereas volume of a sphere is proportional to the third power of its radius, light flux decreases only as the inverse square power with distance from its source. Calculation therefore shows that with an infinite volume of stars the sky would everywhere present the light intensity of the average star surface intensity.

"This is reassuring to me," said Peter. "Each time I go out at night I can look to the sky and see that vast as the universe is there is a finitude about it."

"Certainly the darkness of the night sky and Olbers' calculation," Roberto continued, "constitute one of the key principles which must guide any cosmology. And it is nice that this basic cosmological fact is so simply and readily open to observation by all. However, remember that Olbers' consideration gives no guarantee of a finite spatial universe. All we can say is that stars or galaxies cannot extend indefinitely with about the same density that we find in our observable universe. Either the density must fall off, or there is the possibility envisaged by the steady-state theory, that the galaxies that are sufficiently far away are receding with velocities so close to that of light that the amount of light energy received from them becomes negligible. But let us assume that we do have a universe of a finite number of galaxies in a finite volume. Then the expansion of the universe points to an infinity in that the expansion may be carrying space and the galaxies outward forever. We come then to an infinite spatial extension. However, Peter, there is an alternative that might keep us even from this infinity. One of the model universes which cosmological theorists have derived from the general relativity equations gives the property that the expansion of the universe will in time decelerate, and eventually cease. In the next phase, the process will begin to go the other way: there will be contraction, and the galaxies will generally move toward each other rather than apart. But after billions of years this process will bring the universe to a state of density such that expansion will start again. And so on. On this model then our universe is an oscillating one, its vast extent expanding and contracting in great cycles."

"The idea that the universe should have such an overall cyclic behavior is very appealing to me," Peter said. "The recurring cycles bring an atemporal quality to the very cosmos itself. But tell me, is the oscillating-model hypothesis anything more than an idle speculation?"

"Most of cosmology is rather speculative," Roberto replied, "but the oscillating-universe model is one of those that are based on the relevant fact and theory that we do have, and there is as much justification for it as for any of the alternative models. As I said, perhaps the astronomers will soon be giving us more definite evidence for or against any one cosmological model."

Roberto's last words had the effect of so placing the problem of cosmology in observational astronomy that further discussion of it right then seemed fruitless. There was a short silence.

Edwin had been listening to Roberto and Peter with interest, but he seemed also to be following a train of thought of his own. He now spoke. "The existence of the dark night sky and the galactic recessional motions, as two key cosmological facts, is interesting indeed. I think though that there is another fact that is equally impressive, particularly because of its role in giving a progressive, non-cyclical aspect of change to the universe that we know. I refer to the increase in order that we find, in both the living and non-living parts of the world.

"Think," Edwin went on, "of our own galaxy. Presumably there was an original vast cloud of matter particles from which the stars condensed, with formation in at least one case of a planetary system, and in all likelihood in many, many cases. Here then is a process of a chaotic system's taking on the order that is associated with spatial localization of matter. The earth itself perhaps began as a kind of cloud of matter, with eventual condensation and stratification into our globe. All this in the inorganic realm—and in that realm perhaps we should see the formation of the atoms as results of an order-increasing process. Very possibly, as some cosmologists aver, in its early stages of development our galactic universe was constituted only of elementary particles. The formation of nuclei out of neutrons and protons, and of atoms out of nuclei and electrons, and then of molecules out of atoms: these are all processes that bring increased order out of previous random mixing. But once the inorganic world was formed, more or less as we know it today, we know that a tremendous further development of non-random arranging took place in the development of biological systems."

At this point Roberto interposed a question. "Are you thinking," he asked, "of the kind of development that occurred in our own galaxy as perhaps now going on elsewhere in the universe, or are you implying an original chaotic state, as it were, at more or less the same time for all parts of the universe?"

"Oh, my concern is just with the order that we can see and can infer to have developed from previous less ordered states. But as a hypothesis I would take the notion of a much more condensed and much less ordered state, for the entire universe, some several billion years ago. The remarkable thing," Edwin continued, "is the clear fact of increasing complexity in the natural world. We are particularly certain about this fact in the realm of biology. The division between the animate and inanimate world is not one that can be made with any definiteness, but perhaps the first things that could be called living things were molecules that had both a property of forming other molecules in their likeness and also some helpful properties of response to their surroundings. It's a long stride to the single-cell organism which can absorb food, and convert it to its purposes, and reproduce itself. This is a far more complex organism than a single molecule, for already in the single-cell animal there is an organization of billions of individual molecules. Now we know that as evolution proceeded there was development from single-celled to the almost incredibly complex multi-celled plants and animals that have appeared. There have been false starts and mistakes, and now and then nature appears to go awry with some one individual organism, but viewed in the large the development of the species, in the period of a few hundred million years, is a process of steady and sure formation of new systems, securely formed and of greater complexity than preceding ones. This increase in coherence and complexity of pattern is over the relatively simple or even random arrangements of atoms in the inanimate world. In man evolutionary development has come to a biological organism that comprises some 10^{13} cells that function harmoniously, and the organism possesses a brain which allows man his rich life of thought and awareness."

"You almost seem," commented Peter, "to be expressing a view in favor of purposive evolution, or of a teleology operating in nature."

"No, I did not intend that, at least not in any obvious way. But I do want to emphasize the fact of progressive change, perhaps in part because of the emphasis you have put on the cyclical changes. The point is, nature does have both aspects. The time-reversible processes do give rise to evolutionary, irreversible development, and that development gives a basis for past to future time direction; and also, it presents the question of how one should extrapolate the development into future states of the universe, just as there is the question of extrapolation with respect to the expansion of the universe."

"It does seem to me," Peter said, "that the development of the mind is the highest reach of biological evolution so far. Certainly extrapolation would be along the lines of further projections of our cognitive life."

"Yes, I think that is right," Edwin replied. "The development of mind is the culmination that we can discern in the drift toward states of increased complexity of pattern. And we have generally come at least to the wisdom that mind is part of the natural order. I think there is no level in the hierarchy of complex systems at which one can say mind begins. Even, say, among the atoms combining to form complex molecules there is, I should think, a response to environment that could be called rudimentary awareness. And as we go through the various orders of living things we are hard put to say that it is in any one that mind begins. Organisms with minds will surely develop in the universe, given suitable physical conditions for the existence of sufficiently complex systems. One extrapolation which I would think we could make is that there will be communication, and maybe in time even some kind of symbiosis, between minds in different parts of the universe. I would think too that we could expect increasing amounts of activities that are directed by understanding, or science, rather than by instinct or tradition. We see a beginning of this guidance in mankind's affairs: results of science have affected many of his living habits, and certainly man has consciously changed the surface of the earth—or unfortunately despoiled it—as has no other living thing."

Roberto entered the conversation again to say, "I agree with you; it is striking that we find such development of complex systems both in the inorganic and in the organic realms. Now, we do have the general principle that in an isolated system there is entropy increase as time goes on, and that means increase in disorder or loss of pattern in favor of random arrangement. Do you assume that an increase of ordered arrangements in our galaxy, say, is compensated for by disorder increase elsewhere in the universe?"

"I don't know the answer to the problem you raise," Edwin said. "I am prepared to stand by the principle of entropy increase for any isolated system. And, as we discussed earlier, entropy decrease for any one organism can be achieved at the expense of the surroundings, so that there is a net entropy increase, in accordance with the principle. With respect to the entropy of galactic systems, I suppose we would find if we studied the matter closely that entropy decrease could likewise be at the expense of the surrounding universe. I don't know that any one has made any attempt carefully to define entropy, or to apply the *H*-theorem, for galaxies; certainly it would not be a simple matter to do so. But, I would think that there is a possibility that the expansion of the universe does itself provide an entropy sink, as it were. The expansion might be a cooling, thermally speaking, that allows for a general decrease of entropy in the entire universe, and hence favors the formation of constituent systems of lower entropy and smaller probability of formation."

Edwin paused a second, and then continued. "We do after all have the problem of why systems of higher degree of order than their surroundings do emerge. We can describe the processes of emergence, in terms of our physico-chemical laws, and yet there is still some question, it seems to me, about why nature is so constituted that the emergence occurs. We touched on this before: why do the various time-reversible processes at an elementary level combine to give us, not only irreversible processes, but in some cases irreversible processes that result in part of the total system having a higher degree of order—a lower entropy—than did any part of the system originally. Think of a glass of water whose temperature is falling. When it comes to zero degrees Centigrade, the water gives up a relatively large amount of heat and freezes. That is, the water molecules fall into the orderly crystalline arrangement of the ice lattice. Now, what was required for the entropy decrease or increase of order on freezing was that heat could be radiated from the cold water. The surroundings had to be colder than the water. Suppose we consider living things—inorganic substances and sunlight give the material bases for their formation. But similarly, I would think it could be argued that the entropy decrease that is associated with the formation of biological systems over the surface of the earth occurs because there can be radiation into surrounding cooler space; likewise too, perhaps, with the development processes in galaxies and stars. It is just possible, then, that the forming of ordered systems from relatively less ordered surroundings is associated with the forming of empty space, as the universe expands, into which energy may be radiated."

Peter said, "Then we have a guarantee in the expansion of the universe that for a long time to come conditions will favor the development of complex systems. And surely this does mean the continued growth of man's mental capacities?"

"Yes," said Edwin, "if we wish to accept my suggestion, we can foresee a long development of intellectual life, for it does seem to be that deliberate study of the natural world, as we now see in man, does come with biological systems of high order. We must remember too, that information itself represents entropy decrease, or, as it has come to be called, negentropy or negative entropy. Communication engineers have found this association between entropy and information to be very useful. An electrical communications system, say, which is giving only the random signals that are inherent in the molecular motions of the system might be said to be at a base-zero entropy. The sending of an intelligible signal through the system—say just a coherent set of pips—is an imposing of order on the system, and may be represented as a decrease of entropy. As you probably know, ingenious work has been done in analyzing sentences of ordinary discourse for their information content, in negentropy terms. And the concept of negentropy has been nicely brought into consonance with the entropy concept of statistical mechanics.

"Mankind has made only a small beginning," Edwin continued, "but in view of the solid facts of evolutionary development, it seems almost certain that there will be continued development of capacity for understanding and awareness in biological organisms. I think also that the scientific discerning of facts and the elaboration of instruments for that discerning, will be increasingly successful. True, at the moment it appears that man may destroy his species rather than carry on the fine start that he has made, but if not man and his descendants then I think some other organism will continue the development of awareness of the universe. Indeed, who knows how far this process may have already gone elsewhere in the galaxy. But like all of us, I hope that we—man—will achieve intelligent guidance of our social affairs, even for millions of years, and be able to carry on the enterprise of the sciences and the arts. I would expect that man in the future could reach an understanding of himself and the universe that is almost incredible to us. We might expect that there would be an availability of any desired fact, and also of theory of any desired order of generality for the bringing of coherence among facts."

"Do you think," asked Peter, "of this development as continuing indefinitely?"

"No, I would not think so," Edwin replied. "There surely would be natural limits to the degree of organization which could obtain among the entities of any one system—among the molecules of the earth's crust, to take an extreme example. We might perhaps think that the descendants of man would have eventually reached a state of individual and social organization—and of awareness—such that no further change is desirable, and equilibrium has been protected."

Roberto slightly raised his hand, to indicate that he wished to say something. "Your association, Edwin," he said, "of increase of organization with a state of expansion of the universe is appealing. I wonder, though, if a decrease of energy density in the universe won't eventually give you too much of what you want. If the sun, say, were past its period of energy production, and the solar system continued to receive no more energy from the other stars than it now does, temperatures here would approach very close to absolute zero. Certainly you do not want, in your final state, a complete ordering of the kind that can be achieved at absolute zero—for in that ordering there is a minimum of the complexity of matter that is to be associated with a highly developed mind."

"Your point is a good one," Edwin replied, "yet I think it need not invalidate my proposal. Speaking in the speculative way that we are, we can assume that man will be able to overcome great problems in the control of the physical world for his needs. I could not venture any predictions about the state of our galaxy hundreds of millions, or even billions, of years hence; but I should think that the man of that day could take advantage of whatever possibilities exist, and could even move to parts of the galaxy far removed from the earth if necessary."

After a pause Edwin continued, "Your comment, Roberto, does suggest that a period in which there is development of highly organized systems—like the period we of the earth are part of—does require both a high energy source which gives rise to change, to possibilities for new arrangements of atoms, and also an environment that allows for entropy loss and stability. Our sun provides energy for changing states, and the relatively dark night sky allows a contrasting cooling and stability. Very likely a terminal state of low and uniform energy density in the universe would be one in which there could be no further development in complexity. But in the state that I am proposing there would, after all, be no desire or expectation of further change. I would think that the complex organizations already achieved could be maintained. There would be no need for any processes of entropy change, and complex but enduring patterns of cyclic process on the atomic level might alone suffice for the ultimate processes of contemplation and awareness."

Peter made the comment, "It is curious to think what this state might be. There would appear to be a rather dull finality about it."

"No, on the contrary," Edwin replied emphatically. "We should think perhaps not only physically but also in terms of the conscious life that will be enjoyed by our highly complex organisms. There will be the great achievements of the past to recall—for we can be sure that the state I suggest will not be gained without the work of what we would call heroes and great men. The understanding of the universe will be a source of satisfaction, and I think that our future man will not have devised an ultimate state that leaves him prey to pain and unhappiness."

"I am reminded," said Peter, "of Aristotle's Prime Mover. Our ultimate man would, I presume, lead essentially a life of contemplation of the universe. I would rephrase your suggestion, Edwin, about his enjoying the achievements of the past, and say that in man's final state the valuable things that he has achieved in his long history are to be preserved: not only his scientific understanding, but his aesthetic awareness, perhaps even some of his artefacts, and certainly his sense of regard and respect for fellow sentient beings. In a word, what we strive for today is not 'mere sound and fury, signifying nothing,' but something that will be measured and hallowed in the life of far days to come."

Edwin again took up the discussion. "It is interesting to think about how awareness of the past is preserved. This is tantamount I suppose to asking how is memory possible. One theory has been that a closed neuron circuit in the brain corresponds to a memory. The circuit remains more or less indefinitely, and can be brought to awareness when called up as a result of the remarkable sense of relevance that man has. An alternative theory which I recently heard about is that each memory corresponds to an organic molecule in the brain; the many, many possibilities for variation in structure among fairly complex molecules then give a basis for our richness of memory. In any case, I think that some physical process must be associated with memory. It occurred to me earlier when we were discussing time-reversible processes and enduring systems, that memory must be associated with a kind of persistent system that is in some way out of the stream of progressively changing processes. With respect, then, to the final and complex systems that we have been talking about as the culmination of evolution, I should like to think that they will consist of persistent physical processes, with only the cyclical changes of a stable system and in a universe that is essentially in equilibrium. These processes would be the physical concomitant of the contemplation of the past and present of the universe."

"Ah," said Peter, "we then come in the end to a time-less universe after all. The progressive development that has provided a time direction brings us eventually to a universe of equilibrium cyclic processes, and with awareness of itself and its past. I suppose we could say, then, that we have come to the end of time. It would be a proper prize indeed for the universe to win through to this final state of enjoyment."

Roberto smiled, and said, "But I feel that we, here today, have now carried speculation to the limit. The cosmology we have come to is a satisfactory one, and seems to me to be in the nature of a final speculation."

Peter, however, held up his hand, to indicate that he had something more to add. "I very much like the notion of an eventual time-less and reflective universe," he said, "but as an honest philosopher I must confess that I am not satisfied with the idea. It is well enough for us three professors to envision an end state of placid understanding and awareness, but life around us everywhere points also to other aspects of nature. There are activity, force, passion and destruction as well as contemplation and understanding. Could a universe with all these tendencies eventually become one only of quiet reflection?"

For a few seconds there was no response to Peter's question, but then Roberto spoke, with a twinkle in his eye. "I think you may be right, Peter, but we need not give up the course of development that has seemed plausible to us. Remember the oscillating-model universe. Let us accept that; then the final world which Edwin suggested will not last indefinitely. The contraction phase will eventually start, at first occurring slowly but gradually increasing in pace. Greater amounts of radiant energy will come to the various galactic systems; in time there will be increase everywhere in disorder and in destruction of complex systems, as the universe moves towards a highly condensed, high temperature state. The whole vast drama that we have been discussing will now have to repeat itself—possibly with some variations."

"Well done, Roberto!" Peter said. "And now I don't believe that I have anything more to add."

But Edwin did add, "Let us hope that in the state of contemplation our ultimate man does understand why the repetition is necessary."

Peter sat up, and announced, "The children and ladies are back from the beach, and a time ago my wife made a signal through the window that tea would soon be ready. What do you say, gentlemen, shall we join the ladies inside?"

"That would be most pleasant," said Edwin.

"Yes, it would," Roberto concurred. "Our discussion has been a good one, but now it is fitting to return to the ordinary affairs of the day."

"I think too," said Peter, "that our wives will be interested in some of the things that we have said."

The three men arose and passed from the porch into the cottage.

Index

SOME DOVER SCIENCE BOOKS

SOME DOVER SCIENCE BOOKS

WHAT IS SCIENCE?,
Norman Campbell

This excellent introduction explains scientific method, role of mathematics, types of scientific laws. Contents: 2 aspects of science, science & nature, laws of science, discovery of laws, explanation of laws, measurement & numerical laws, applications of science. 192pp. $5\frac{3}{8}$ x 8. Paperbound $1.25

FADS AND FALLACIES IN THE NAME OF SCIENCE,
Martin Gardner

Examines various cults, quack systems, frauds, delusions which at various times have masqueraded as science. Accounts of hollow-earth fanatics like Symmes; Velikovsky and wandering planets; Hoerbiger; Bellamy and the theory of multiple moons; Charles Fort; dowsing, pseudoscientific methods for finding water, ores, oil. Sections on naturopathy, iridiagnosis, zone therapy, food fads, etc. Analytical accounts of Wilhelm Reich and orgone sex energy; L. Ron Hubbard and Dianetics; A. Korzybski and General Semantics; many others. Brought up to date to include Bridey Murphy, others. Not just a collection of anecdotes, but a fair, reasoned appraisal of eccentric theory. Formerly titled *In the Name of Science*. Preface. Index. x + 384pp. $5\frac{3}{8}$ x 8.

Paperbound $1.85

PHYSICS, THE PIONEER SCIENCE,
L. W. Taylor

First thorough text to place all important physical phenomena in cultural-historical framework; remains best work of its kind. Exposition of physical laws, theories developed chronologically, with great historical, illustrative experiments diagrammed, described, worked out mathematically. Excellent physics text for self-study as well as class work. Vol. 1: Heat, Sound: motion, acceleration, gravitation, conservation of energy, heat engines, rotation, heat, mechanical energy, etc. 211 illus. 407pp. $5\frac{3}{8}$ x 8. Vol. 2: Light, Electricity: images, lenses, prisms, magnetism, Ohm's law, dynamos, telegraph, quantum theory, decline of mechanical view of nature, etc. Bibliography. 13 table appendix. Index. 551 illus. 2 color plates. 508pp. $5\frac{3}{8}$ x 8.

Vol. 1 Paperbound $2.25, Vol. 2 Paperbound $2.25,
The set $4.50

THE EVOLUTION OF SCIENTIFIC THOUGHT FROM NEWTON TO EINSTEIN,
A. d'Abro

Einstein's special and general theories of relativity, with their historical implications, are analyzed in non-technical terms. Excellent accounts of the contributions of Newton, Riemann, Weyl, Planck, Eddington, Maxwell, Lorentz and others are treated in terms of space and time, equations of electromagnetics, finiteness of the universe, methodology of science. 21 diagrams. 482pp. $5\frac{3}{8}$ x 8.

Paperbound $2.50

CHANCE, LUCK AND STATISTICS: THE SCIENCE OF CHANCE,
Horace C. Levinson

Theory of probability and science of statistics in simple, non-technical language. Part I deals with theory of probability, covering odd superstitions in regard to "luck," the meaning of betting odds, the law of mathematical expectation, gambling, and applications in poker, roulette, lotteries, dice, bridge, and other games of chance. Part II discusses the misuse of statistics, the concept of statistical probabilities, normal and skew frequency distributions, and statistics applied to various fields—birth rates, stock speculation, insurance rates, advertising, etc. "Presented in an easy humorous style which I consider the best kind of expository writing," Prof. A. C. Cohen, Industry Quality Control. Enlarged revised edition. Formerly titled *The Science of Chance*. Preface and two new appendices by the author. Index. xiv + 365pp. 5⅜ x 8. Paperbound $2.00

BASIC ELECTRONICS,
prepared by the U.S. Navy Training Publications Center

A thorough and comprehensive manual on the fundamentals of electronics. Written clearly, it is equally useful for self-study or course work for those with a knowledge of the principles of basic electricity. Partial contents: Operating Principles of the Electron Tube; Introduction to Transistors; Power Supplies for Electronic Equipment; Tuned Circuits; Electron-Tube Amplifiers; Audio Power Amplifiers; Oscillators; Transmitters; Transmission Lines; Antennas and Propagation; Introduction to Computers; and related topics. Appendix. Index. Hundreds of illustrations and diagrams. vi + 471pp. 6½ x 9¼.
Paperbound $2.75

BASIC THEORY AND APPLICATION OF TRANSISTORS,
prepared by the U.S. Department of the Army

An introductory manual prepared for an army training program. One of the finest available surveys of theory and application of transistor design and operation. Minimal knowledge of physics and theory of electron tubes required. Suitable for textbook use, course supplement, or home study. Chapters: Introduction; fundamental theory of transistors; transistor amplifier fundamentals; parameters, equivalent circuits, and characteristic curves; bias stabilization; transistor analysis and comparison using characteristic curves and charts; audio amplifiers; tuned amplifiers; wide-band amplifiers; oscillators; pulse and switching circuits; modulation, mixing, and demodulation; and additional semiconductor devices. Unabridged, corrected edition. 240 schematic drawings, photographs, wiring diagrams, etc. 2 Appendices. Glossary. Index. 263pp. 6½ x 9¼. Paperbound $1.25

GUIDE TO THE LITERATURE OF MATHEMATICS AND PHYSICS,
N. G. Parke III

Over 5000 entries included under approximately 120 major subject headings of selected most important books, monographs, periodicals, articles in English, plus important works in German, French, Italian, Spanish, Russian (many recently available works). Covers every branch of physics, math, related engineering. Includes author, title, edition, publisher, place, date, number of volumes, number of pages. A 40-page introduction on the basic problems of research and study provides useful information on the organization and use of libraries, the psychology of learning, etc. This reference work will save you hours of time. 2nd revised edition. Indices of authors, subjects, 464pp. 5⅜ x 8.
Paperbound $2.75

THE RISE OF THE NEW PHYSICS (formerly THE DECLINE OF MECHANISM), *A. d'Abro*

This authoritative and comprehensive 2-volume exposition is unique in scientific publishing. Written for intelligent readers not familiar with higher mathematics, it is the only thorough explanation in non-technical language of modern mathematical-physical theory. Combining both history and exposition, it ranges from classical Newtonian concepts up through the electronic theories of Dirac and Heisenberg, the statistical mechanics of Fermi, and Einstein's relativity theories. "A must for anyone doing serious study in the physical sciences," *J. of Franklin Inst.* 97 illustrations. 991pp. 2 volumes.

Vol. 1 Paperbound $2.25, Vol. 2 Paperbound $2.25, The set $4.50

THE STRANGE STORY OF THE QUANTUM, AN ACCOUNT FOR THE GENERAL READER OF THE GROWTH OF IDEAS UNDERLYING OUR PRESENT ATOMIC KNOWLEDGE, *B. Hoffmann*

Presents lucidly and expertly, with barest amount of mathematics, the problems and theories which led to modern quantum physics. Dr. Hoffmann begins with the closing years of the 19th century, when certain trifling discrepancies were noticed, and with illuminating analogies and examples takes you through the brilliant concepts of Planck, Einstein, Pauli, de Broglie, Bohr, Schroedinger, Heisenberg, Dirac, Sommerfeld, Feynman, etc. This edition includes a new, long postscript carrying the story through 1958. "Of the books attempting an account of the history and contents of our modern atomic physics which have come to my attention, this is the best," H. Margenau, Yale University, in *American Journal of Physics.* 32 tables and line illustrations. Index. 275pp. 5⅜ x 8.

Paperbound $1.75

GREAT IDEAS AND THEORIES OF MODERN COSMOLOGY, *Jagjit Singh*

The theories of Jeans, Eddington, Milne, Kant, Bondi, Gold, Newton, Einstein, Gamow, Hoyle, Dirac, Kuiper, Hubble, Weizsäcker and many others on such cosmological questions as the origin of the universe, space and time, planet formation, "continuous creation," the birth, life, and death of the stars, the origin of the galaxies, etc. By the author of the popular *Great Ideas of Modern Mathematics.* A gifted popularizer of science, he makes the most difficult abstractions crystal-clear even to the most non-mathematical reader. Index. xii + 276pp. 5⅜ x 8½. Paperbound $2.00

GREAT IDEAS OF MODERN MATHEMATICS: THEIR NATURE AND USE, *Jagjit Singh*

Reader with only high school math will understand main mathematical ideas of modern physics, astronomy, genetics, psychology, evolution, etc., better than many who use them as tools, but comprehend little of their basic structure. Author uses his wide knowledge of non-mathematical fields in brilliant exposition of differential equations, matrices, group theory, logic, statistics, problems of mathematical foundations, imaginary numbers, vectors, etc. Original publications, 2 appendices. 2 indexes. 65 illustr. 322pp. 5⅜ x 8. Paperbound $2.00

THE MATHEMATICS OF GREAT AMATEURS, *Julian L. Coolidge*

Great discoveries made by poets, theologians, philosophers, artists and other non-mathematicians: Omar Khayyam, Leonardo da Vinci, Albrecht Dürer, John Napier, Pascal, Diderot, Bolzano, etc. Surprising accounts of what can result from a non-professional preoccupation with the oldest of sciences. 56 figures. viii + 211pp. 5⅜ x 8½. Paperbound $1.50

COLLEGE ALGEBRA, *H. B. Fine*
Standard college text that gives a systematic and deductive structure to algebra; comprehensive, connected, with emphasis on theory. Discusses the commutative, associative, and distributive laws of number in unusual detail, and goes on with undetermined coefficients, quadratic equations, progressions, logarithms, permutations, probability, power series, and much more. Still most valuable elementary-intermediate text on the science and structure of algebra. Index. 1560 problems, all with answers. x + 631pp. 5⅜ x 8. Paperbound $2.75

HIGHER MATHEMATICS FOR STUDENTS OF CHEMISTRY AND PHYSICS, *J. W. Mellor*
Not abstract, but practical, building its problems out of familiar laboratory material, this covers differential calculus, coordinate, analytical geometry, functions, integral calculus, infinite series, numerical equations, differential equations, Fourier's theorem, probability, theory of errors, calculus of variations, determinants. "If the reader is not familiar with this book, it will repay him to examine it," *Chem. & Engineering News.* 800 problems. 189 figures. Bibliography. xxi + 641pp. 5⅜ x 8. Paperbound $2.50

TRIGONOMETRY REFRESHER FOR TECHNICAL MEN, *A. A. Klaf*
A modern question and answer text on plane and spherical trigonometry. Part I covers plane trigonometry: angles, quadrants, trigonometrical functions, graphical representation, interpolation, equations, logarithms, solution of triangles, slide rules, etc. Part II discusses applications to navigation, surveying, elasticity, architecture, and engineering. Small angles, periodic functions, vectors, polar coordinates, De Moivre's theorem, fully covered. Part III is devoted to spherical trigonometry and the solution of spherical triangles, with applications to terrestrial and astronomical problems. Special time-savers for numerical calculation. 913 questions answered for you! 1738 problems; answers to odd numbers. 494 figures. 14 pages of functions, formulae. Index. x + 629pp. 5⅜ x 8. Paperbound $2.00

CALCULUS REFRESHER FOR TECHNICAL MEN, *A. A. Klaf*
Not an ordinary textbook but a unique refresher for engineers, technicians, and students. An examination of the most important aspects of differential and integral calculus by means of 756 key questions. Part I covers simple differential calculus: constants, variables, functions, increments, derivatives, logarithms, curvature, etc. Part II treats fundamental concepts of integration: inspection, substitution, transformation, reduction, areas and volumes, mean value, successive and partial integration, double and triple integration. Stresses practical aspects! A 50 page section gives applications to civil and nautical engineering, electricity, stress and strain, elasticity, industrial engineering, and similar fields. 756 questions answered. 556 problems; solutions to odd numbers. 36 pages of constants, formulae. Index. v + 431pp. 5⅜ x 8. Paperbound $2.00

INTRODUCTION TO THE THEORY OF GROUPS OF FINITE ORDER, *R. Carmichael*
Examines fundamental theorems and their application. Beginning with sets, systems, permutations, etc., it progresses in easy stages through important types of groups: Abelian, prime power, permutation, etc. Except 1 chapter where matrices are desirable, no higher math needed. 783 exercises, problems. Index. xvi + 447pp. 5⅜ x 8. Paperbound $3.00

FIVE VOLUME "THEORY OF FUNCTIONS" SET BY KONRAD KNOPP

This five-volume set, prepared by Konrad Knopp, provides a complete and readily followed account of theory of functions. Proofs are given concisely, yet without sacrifice of completeness or rigor. These volumes are used as texts by such universities as M.I.T., University of Chicago, N. Y. City College, and many others. "Excellent introduction . . . remarkably readable, concise, clear, rigorous," *Journal of the American Statistical Association.*

ELEMENTS OF THE THEORY OF FUNCTIONS,
Konrad Knopp
This book provides the student with background for further volumes in this set, or texts on a similar level. Partial contents: foundations, system of complex numbers and the Gaussian plane of numbers, Riemann sphere of numbers, mapping by linear functions, normal forms, the logarithm, the cyclometric functions and binomial series. "Not only for the young student, but also for the student who knows all about what is in it," *Mathematical Journal.* Bibliography. Index. 140pp. 5⅜ x 8. Paperbound $1.50

THEORY OF FUNCTIONS, PART I,
Konrad Knopp
With volume II, this book provides coverage of basic concepts and theorems. Partial contents: numbers and points, functions of a complex variable, integral of a continuous function, Cauchy's integral theorem, Cauchy's integral formulae, series with variable terms, expansion of analytic functions in power series, analytic continuation and complete definition of analytic functions, entire transcendental functions, Laurent expansion, types of singularities. Bibliography. Index. vii + 146pp. 5⅜ x 8. Paperbound $1.35

THEORY OF FUNCTIONS, PART II,
Konrad Knopp
Application and further development of general theory, special topics. Single valued functions. Entire, Weierstrass, Meromorphic functions. Riemann surfaces. Algebraic functions. Analytical configuration, Riemann surface. Bibliography. Index. x + 150pp. 5⅜ x 8. Paperbound $1.35

PROBLEM BOOK IN THE THEORY OF FUNCTIONS, VOLUME 1.
Konrad Knopp
Problems in elementary theory, for use with Knopp's *Theory of Functions,* or any other text, arranged according to increasing difficulty. Fundamental concepts, sequences of numbers and infinite series, complex variable, integral theorems, development in series, conformal mapping. 182 problems. Answers. viii + 126pp. 5⅜ x 8. Paperbound $1.35

PROBLEM BOOK IN THE THEORY OF FUNCTIONS, VOLUME 2,
Konrad Knopp
Advanced theory of functions, to be used either with Knopp's *Theory of Functions,* or any other comparable text. Singularities, entire & meromorphic functions, periodic, analytic, continuation, multiple-valued functions, Riemann surfaces, conformal mapping. Includes a section of additional elementary problems. "The difficult task of selecting from the immense material of the modern theory of functions the problems just within the reach of the beginner is here masterfully accomplished," *Am. Math. Soc.* Answers. 138pp. 5⅜ x 8. Paperbound $1.50

NUMERICAL SOLUTIONS OF DIFFERENTIAL EQUATIONS,
H. Levy & E. A. Baggott

Comprehensive collection of methods for solving ordinary differential equations of first and higher order. All must pass 2 requirements: easy to grasp and practical, more rapid than school methods. Partial contents: graphical integration of differential equations, graphical methods for detailed solution. Numerical solution. Simultaneous equations and equations of 2nd and higher orders. "Should be in the hands of all in research in applied mathematics, teaching," *Nature*. 21 figures. viii + 238pp. 5⅜ x 8. Paperbound $1.85

ELEMENTARY STATISTICS, WITH APPLICATIONS IN MEDICINE AND THE BIOLOGICAL SCIENCES, *F. E. Croxton*

A sound introduction to statistics for anyone in the physical sciences, assuming no prior acquaintance and requiring only a modest knowledge of math. All basic formulas carefully explained and illustrated; all necessary reference tables included. From basic terms and concepts, the study proceeds to frequency distribution, linear, non-linear, and multiple correlation, skewness, kurtosis, etc. A large section deals with reliability and significance of statistical methods. Containing concrete examples from medicine and biology, this book will prove unusually helpful to workers in those fields who increasingly must evaluate, check, and interpret statistics. Formerly titled "Elementary Statistics with Applications in Medicine." 101 charts. 57 tables. 14 appendices. Index. vi + 376pp. 5⅜ x 8. Paperbound $2.00

INTRODUCTION TO SYMBOLIC LOGIC,
S. Langer

No special knowledge of math required — probably the clearest book ever written on symbolic logic, suitable for the layman, general scientist, and philosopher. You start with simple symbols and advance to a knowledge of the Boole-Schroeder and Russell-Whitehead systems. Forms, logical structure, classes, the calculus of propositions, logic of the syllogism, etc. are all covered. "One of the clearest and simplest introductions," *Mathematics Gazette*. Second enlarged, revised edition. 368pp. 5⅜ x 8. Paperbound $2.00

A SHORT ACCOUNT OF THE HISTORY OF MATHEMATICS,
W. W. R. Ball

Most readable non-technical history of mathematics treats lives, discoveries of every important figure from Egyptian, Phoenician, mathematicians to late 19th century. Discusses schools of Ionia, Pythagoras, Athens, Cyzicus, Alexandria, Byzantium, systems of numeration; primitive arithmetic; Middle Ages, Renaissance, including Arabs, Bacon, Regiomontanus, Tartaglia, Cardan, Stevinus, Galileo, Kepler; modern mathematics of Descartes, Pascal, Wallis, Huygens, Newton, Leibnitz, d'Alembert, Euler, Lambert, Laplace, Legendre, Gauss, Hermite, Weierstrass, scores more. Index. 25 figures. 546pp. 5⅜ x 8. Paperbound $2.25

INTRODUCTION TO NONLINEAR DIFFERENTIAL AND INTEGRAL EQUATIONS,
Harold T. Davis

Aspects of the problem of nonlinear equations, transformations that lead to equations solvable by classical means, results in special cases, and useful generalizations. Thorough, but easily followed by mathematically sophisticated reader who knows little about non-linear equations. 137 problems for student to solve. xv + 566pp. 5⅜ x 8½. Paperbound $2.00

AN INTRODUCTION TO THE GEOMETRY OF N DIMENSIONS, *D. H. Y. Sommerville*

An introduction presupposing no prior knowledge of the field, the only book in English devoted exclusively to higher dimensional geometry. Discusses fundamental ideas of incidence, parallelism, perpendicularity, angles between linear space; enumerative geometry; analytical geometry from projective and metric points of view; polytopes; elementary ideas in analysis situs; content of hyper-spacial figures. Bibliography. Index. 60 diagrams. 196pp. 5⅜ x 8.
Paperbound $1.50

ELEMENTARY CONCEPTS OF TOPOLOGY, *P. Alexandroff*

First English translation of the famous brief introduction to topology for the beginner or for the mathematician not undertaking extensive study. This unusually useful intuitive approach deals primarily with the concepts of complex, cycle, and homology, and is wholly consistent with current investigations. Ranges from basic concepts of set-theoretic topology to the concept of Betti groups. "Glowing example of harmony between intuition and thought," David Hilbert. Translated by A. E. Farley. Introduction by D. Hilbert. Index. 25 figures. 73pp. 5⅜ x 8. Paperbound $1.00

ELEMENTS OF NON-EUCLIDEAN GEOMETRY, *D. M. Y. Sommerville*

Unique in proceeding step-by-step, in the manner of traditional geometry. Enables the student with only a good knowledge of high school algebra and geometry to grasp elementary hyperbolic, elliptic, analytic non-Euclidean geometries; space curvature and its philosophical implications; theory of radical axes; homothetic centres and systems of circles; parataxy and parallelism; absolute measure; Gauss' proof of the defect area theorem; geodesic representation; much more, all with exceptional clarity. 126 problems at chapter endings provide progressive practice and familiarity. 133 figures. Index. xvi + 274pp. 5⅜ x 8. Paperbound $2.00

INTRODUCTION TO THE THEORY OF NUMBERS, *L. E. Dickson*

Thorough, comprehensive approach with adequate coverage of classical literature, an introductory volume beginners can follow. Chapters on divisibility, congruences, quadratic residues & reciprocity. Diophantine equations, etc. Full treatment of binary quadratic forms without usual restriction to integral coefficients. Covers infinitude of primes, least residues. Fermat's theorem. Euler's phi function, Legendre's symbol, Gauss's lemma, automorphs, reduced forms, recent theorems of Thue & Siegel, many more. Much material not readily available elsewhere. 239 problems. Index. I figure. viii + 183pp. 5⅜ x 8.
Paperbound $1.75

MATHEMATICAL TABLES AND FORMULAS, *compiled by Robert D. Carmichael and Edwin R. Smith*

Valuable collection for students, etc. Contains all tables necessary in college algebra and trigonometry, such as five-place common logarithms, logarithmic sines and tangents of small angles, logarithmic trigonometric functions, natural trigonometric functions, four-place antilogarithms, tables for changing from sexagesimal to circular and from circular to sexagesimal measure of angles, etc. Also many tables and formulas not ordinarily accessible, including powers, roots, and reciprocals, exponential and hyperbolic functions, ten-place logarithms of prime numbers, and formulas and theorems from analytical and elementary geometry and from calculus. Explanatory introduction. viii + 269pp. 5⅜ x 8½. Paperbound $1.25

A SOURCE BOOK IN MATHEMATICS,
D. E. Smith
Great discoveries in math, from Renaissance to end of 19th century, in English translation. Read announcements by Dedekind, Gauss, Delamain, Pascal, Fermat, Newton, Abel, Lobachevsky, Bolyai, Riemann, De Moivre, Legendre, Laplace, others of discoveries about imaginary numbers, number congruence, slide rule, equations, symbolism, cubic algebraic equations, non-Euclidean forms of geometry, calculus, function theory, quaternions, etc. Succinct selections from 125 different treatises, articles, most unavailable elsewhere in English. Each article preceded by biographical introduction. Vol. I: Fields of Number, Algebra. Index. 32 illus. 338pp. 5⅜ x 8. Vol. II: Fields of Geometry, Probability, Calculus, Functions, Quaternions. 83 illus. 432pp. 5⅜ x 8.
Vol. 1 Paperbound $2.00, Vol. 2 Paperbound $2.00,
The set $4.00

FOUNDATIONS OF PHYSICS,
R. B. Lindsay & H. Margenau
Excellent bridge between semi-popular works & technical treatises. A discussion of methods of physical description, construction of theory; valuable for physicist with elementary calculus who is interested in ideas that give meaning to data, tools of modern physics. Contents include symbolism; mathematical equations; space & time foundations of mechanics; probability; physics & continua; electron theory; special & general relativity; quantum mechanics; causality. "Thorough and yet not overdetailed. Unreservedly recommended," *Nature* (London). Unabridged, corrected edition. List of recommended readings. 35 illustrations. xi + 537pp. 5⅜ x 8. Paperbound $3.00

FUNDAMENTAL FORMULAS OF PHYSICS,
ed. by D. H. Menzel
High useful, full, inexpensive reference and study text, ranging from simple to highly sophisticated operations. Mathematics integrated into text—each chapter stands as short textbook of field represented. Vol. 1: Statistics, Physical Constants, Special Theory of Relativity, Hydrodynamics, Aerodynamics, Boundary Value Problems in Math, Physics, Viscosity, Electromagnetic Theory, etc. Vol. 2: Sound, Acoustics, Geometrical Optics, Electron Optics, High-Energy Phenomena, Magnetism, Biophysics, much more. Index. Total of 800pp. 5⅜ x 8.
Vol. 1 Paperbound $2.25, Vol. 2 Paperbound $2.25,
The set $4.50

THEORETICAL PHYSICS,
A. S. Kompaneyets
One of the very few thorough studies of the subject in this price range. Provides advanced students with a comprehensive theoretical background. Especially strong on recent experimentation and developments in quantum theory. Contents: Mechanics (Generalized Coordinates, Lagrange's Equation, Collision of Particles, etc.), Electrodynamics (Vector Analysis, Maxwell's equations, Transmission of Signals, Theory of Relativity, etc.), Quantum Mechanics (the Inadequacy of Classical Mechanics, the Wave Equation, Motion in a Central Field, Quantum Theory of Radiation, Quantum Theories of Dispersion and Scattering, etc.), and Statistical Physics (Equilibrium Distribution of Molecules in an Ideal Gas, Boltzmann Statistics, Bose and Fermi Distribution. Thermodynamic Quantities, etc.). Revised to 1961. Translated by George Yankovsky, authorized by Kompaneyets. 137 exercises. 56 figures. 529pp. 5⅜ x 8½.
Paperbound $2.50

MATHEMATICAL PHYSICS, *D. H. Menzel*
Thorough one-volume treatment of the mathematical techniques vital for classical mechanics, electromagnetic theory, quantum theory, and relativity. Written by the Harvard Professor of Astrophysics for junior, senior, and graduate courses, it gives clear explanations of all those aspects of function theory, vectors, matrices, dyadics, tensors, partial differential equations, etc., necessary for the understanding of the various physical theories. Electron theory, relativity, and other topics seldom presented appear here in considerable detail. Scores of definition, conversion factors, dimensional constants, etc. "More detailed than normal for an advanced text . . . excellent set of sections on Dyadics, Matrices, and Tensors," *Journal of the Franklin Institute.* Index. 193 problems, with answers. x + 412pp. 5⅜ x 8. Paperbound $2.50

THE THEORY OF SOUND, *Lord Rayleigh*
Most vibrating systems likely to be encountered in practice can be tackled successfully by the methods set forth by the great Nobel laureate, Lord Rayleigh. Complete coverage of experimental, mathematical aspects of sound theory. Partial contents: Harmonic motions, vibrating systems in general, lateral vibrations of bars, curved plates or shells, applications of Laplace's functions to acoustical problems, fluid friction, plane vortex-sheet, vibrations of solid bodies, etc. This is the first inexpensive edition of this great reference and study work. Bibliography, Historical introduction by R. B. Lindsay. Total of 1040pp. 97 figures. 5⅜ x 8. Vol. 1 Paperbound $2.50, Vol. 2 Paperbound $2.50, The set $5.00

HYDRODYNAMICS, *Horace Lamb*
Internationally famous complete coverage of standard reference work on dynamics of liquids & gases. Fundamental theorems, equations, methods, solutions, background, for classical hydrodynamics. Chapters include Equations of Motion, Integration of Equations in Special Gases, Irrotational Motion, Motion of Liquid in 2 Dimensions, Motion of Solids through Liquid-Dynamical Theory, Vortex Motion, Tidal Waves, Surface Waves, Waves of Expansion, Viscosity, Rotating Masses of Liquids. Excellently planned, arranged; clear, lucid presentation. 6th enlarged, revised edition. Index. Over 900 footnotes, mostly bibliographical. 119 figures. xv + 738pp. 6⅛ x 9¼. Paperbound $4.00

DYNAMICAL THEORY OF GASES, *James Jeans*
Divided into mathematical and physical chapters for the convenience of those not expert in mathematics, this volume discusses the mathematical theory of gas in a steady state, thermodynamics, Boltzmann and Maxwell, kinetic theory, quantum theory, exponentials, etc. 4th enlarged edition, with new material on quantum theory, quantum dynamics, etc. Indexes. 28 figures. 444pp. 6⅛ x 9¼. Paperbound $2.75

THERMODYNAMICS, *Enrico Fermi*
Unabridged reproduction of 1937 edition. Elementary in treatment; remarkable for clarity, organization. Requires no knowledge of advanced math beyond calculus, only familiarity with fundamentals of thermometry, calorimetry. Partial Contents: Thermodynamic systems; First & Second laws of thermodynamics; Entropy; Thermodynamic potentials: phase rule, reversible electric cell; Gaseous reactions: van't Hoff reaction box, principle of LeChatelier; Thermodynamics of dilute solutions: osmotic & vapor pressures, boiling & freezing points; Entropy constant. Index. 25 problems. 24 illustrations. x + 160pp. 5⅜ x 8. Paperbound $1.75

CELESTIAL OBJECTS FOR COMMON TELESCOPES,
Rev. T. W. Webb
Classic handbook for the use and pleasure of the amateur astronomer. Of inestimable aid in locating and identifying thousands of celestial objects. Vol I, The Solar System: discussions of the principle and operation of the telescope, procedures of observations and telescope-photography, spectroscopy, etc., precise location information of sun, moon, planets, meteors. Vol. II, The Stars: alphabetical listing of constellations, information on double stars, clusters, stars with unusual spectra, variables, and nebulae, etc. Nearly 4,000 objects noted. Edited and extensively revised by Margaret W. Mayall, director of the American Assn. of Variable Star Observers. New Index by Mrs. Mayall giving the location of all objects mentioned in the text for Epoch 2000. New Precession Table added. New appendices on the planetary satellites, constellation names and abbreviations, and solar system data. Total of 46 illustrations. Total of xxxix + 606pp. 5⅜ x 8. Vol. 1 Paperbound $2.25, Vol. 2 Paperbound $2.25
The set $4.50

PLANETARY THEORY,
E. W. Brown and C. A. Shook
Provides a clear presentation of basic methods for calculating planetary orbits for today's astronomer. Begins with a careful exposition of specialized mathematical topics essential for handling perturbation theory and then goes on to indicate how most of the previous methods reduce ultimately to two general calculation methods: obtaining expressions either for the coordinates of planetary positions or for the elements which determine the perturbed paths. An example of each is given and worked in detail. Corrected edition. Preface. Appendix. Index. xii + 302pp. 5⅜ x 8½. Paperbound $2.25

STAR NAMES AND THEIR MEANINGS,
Richard Hinckley Allen
An unusual book documenting the various attributions of names to the individual stars over the centuries. Here is a treasure-house of information on a topic not normally delved into even by professional astronomers; provides a fascinating background to the stars in folk-lore, literary references, ancient writings, star catalogs and maps over the centuries. Constellation-by-constellation analysis covers hundreds of stars and other asterisms, including the Pleiades, Hyades, Andromedan Nebula, etc. Introduction. Indices. List of authors and authorities. xx + 563pp. 5⅜ x 8½. Paperbound $2.50

A SHORT HISTORY OF ASTRONOMY, *A. Berry*
Popular standard work for over 50 years, this thorough and accurate volume covers the science from primitive times to the end of the 19th century. After the Greeks and the Middle Ages, individual chapters analyze Copernicus, Brahe, Galileo, Kepler, and Newton, and the mixed reception of their discoveries. Post-Newtonian achievements are then discussed in unusual detail: Halley, Bradley, Lagrange, Laplace, Herschel, Bessel, etc. 2 Indexes. 104 illustrations, 9 portraits. xxxi + 440pp. 5⅜ x 8. Paperbound $2.75

SOME THEORY OF SAMPLING, *W. E. Deming*
The purpose of this book is to make sampling techniques understandable to and useable by social scientists, industrial managers, and natural scientists who are finding statistics increasingly part of their work. Over 200 exercises, plus dozens of actual applications. 61 tables. 90 figs. xix + 602pp. 5⅜ x 8½.
Paperbound $3.50

PRINCIPLES OF STRATIGRAPHY,
A. W. Grabau
Classic of 20th century geology, unmatched in scope and comprehensiveness. Nearly 600 pages cover the structure and origins of every kind of sedimentary, hydrogenic, oceanic, pyroclastic, atmoclastic, hydroclastic, marine hydroclastic, and bioclastic rock; metamorphism; erosion; etc. Includes also the constitution of the atmosphere; morphology of oceans, rivers, glaciers; volcanic activities; faults and earthquakes; and fundamental principles of paleontology (nearly 200 pages). New introduction by Prof. M. Kay, Columbia U. 1277 bibliographical entries. 264 diagrams. Tables, maps, etc. Two volume set. Total of xxxii + 1185pp. 5⅜ x 8. Vol. 1 Paperbound $2.50, Vol. 2 Paperbound $2.50, The set $5.00

SNOW CRYSTALS, *W. A. Bentley and W. J. Humphreys*
Over 200 pages of Bentley's famous microphotographs of snow flakes—the product of painstaking, methodical work at his Jericho, Vermont studio. The pictures, which also include plates of frost, glaze and dew on vegetation, spider webs, windowpanes; sleet; graupel or soft hail, were chosen both for their scientific interest and their aesthetic qualities. The wonder of nature's diversity is exhibited in the intricate, beautiful patterns of the snow flakes. Introductory text by W. J. Humphreys. Selected bibliography. 2,453 illustrations. 224pp. 8 x 10¼. Paperbound $3.25

THE BIRTH AND DEVELOPMENT OF THE GEOLOGICAL SCIENCES,
F. D. Adams
Most thorough history of the earth sciences ever written. Geological thought from earliest times to the end of the 19th century, covering over 300 early thinkers & systems: fossils & their explanation, vulcanists vs. neptunists, figured stones & paleontology, generation of stones, dozens of similar topics. 91 illustrations, including medieval, renaissance woodcuts, etc. Index. 632 footnotes, mostly bibliographical. 511pp. 5⅜ x 8. Paperbound $2.75

ORGANIC CHEMISTRY, *F. C. Whitmore*
The entire subject of organic chemistry for the practicing chemist and the advanced student. Storehouse of facts, theories, processes found elsewhere only in specialized journals. Covers aliphatic compounds (500 pages on the properties and synthetic preparation of hydrocarbons, halides, proteins, ketones, etc.), alicyclic compounds, aromatic compounds, heterocyclic compounds, organophosphorus and organometallic compounds. Methods of synthetic preparation analyzed critically throughout. Includes much of biochemical interest. "The scope of this volume is astonishing," *Industrial and Engineering Chemistry*. 12,000-reference index. 2387-item bibliography. Total of x + 1005pp. 5⅜ x 8. Two volume set, paperbound $4.50

THE PHASE RULE AND ITS APPLICATION,
Alexander Findlay
Covering chemical phenomena of 1, 2, 3, 4, and multiple component systems, this "standard work on the subject" (*Nature*, London), has been completely revised and brought up to date by A. N. Campbell and N. O. Smith. Brand new material has been added on such matters as binary, tertiary liquid equilibria, solid solutions in ternary systems, quinary systems of salts and water. Completely revised to triangular coordinates in ternary systems, clarified graphic representation, solid models, etc. 9th revised edition. Author, subject indexes. 236 figures. 505 footnotes, mostly bibliographic. xii + 494pp. 5⅜ x 8. Paperbound $2.75

A COURSE IN MATHEMATICAL ANALYSIS,
Edouard Goursat
Trans. by E. R. Hedrick, O. Dunkel, H. G. Bergmann. Classic study of fundamental material thoroughly treated. Extremely lucid exposition of wide range of subject matter for student with one year of calculus. Vol. 1: Derivatives and differentials, definite integrals, expansions in series, applications to geometry. 52 figures, 556pp. Paperbound $2.50. Vol. 2, Part 1: Functions of a complex variable, conformal representations, doubly periodic functions, natural boundaries, etc. 38 figures, 269pp. Paperbound $1.85. Vol. 2, Part 2: Differential equations, Cauchy-Lipschitz method, nonlinear differential equations, simultaneous equations, etc. 308pp. Paperbound $1.85. Vol. 3, Part 1: Variation of solutions, partial differential equations of the second order. 15 figures, 339pp. Paperbound $3.00. Vol. 3, Part 2: Integral equations, calculus of variations. 13 figures, 389pp. Paperbound $3.00

PLANETS, STARS AND GALAXIES,
A. E. Fanning
Descriptive astronomy for beginners: the solar system; neighboring galaxies; seasons; quasars; fly-by results from Mars, Venus, Moon; radio astronomy; etc. all simply explained. Revised up to 1966 by author and Prof. D. H. Menzel, former Director, Harvard College Observatory. 29 photos, 16 figures. 189pp. 5⅜ x 8½. Paperbound $1.50

GREAT IDEAS IN INFORMATION THEORY, LANGUAGE AND CYBERNETICS,
Jagjit Singh
Winner of Unesco's Kalinga Prize covers language, metalanguages, analog and digital computers, neural systems, work of McCulloch, Pitts, von Neumann, Turing, other important topics. No advanced mathematics needed, yet a full discussion without compromise or distortion. 118 figures. ix + 338pp. 5⅜ x 8½. Paperbound $2.00

GEOMETRIC EXERCISES IN PAPER FOLDING,
T. Sundara Row
Regular polygons, circles and other curves can be folded or pricked on paper, then used to demonstrate geometric propositions, work out proofs, set up well-known problems. 89 illustrations, photographs of actually folded sheets. xii + 148pp. 5⅜ x 8½. Paperbound $1.00

VISUAL ILLUSIONS, THEIR CAUSES, CHARACTERISTICS AND APPLICATIONS,
M. Luckiesh
The visual process, the structure of the eye, geometric, perspective illusions, influence of angles, illusions of depth and distance, color illusions, lighting effects, illusions in nature, special uses in painting, decoration, architecture, magic, camouflage. New introduction by W. H. Ittleson covers modern developments in this area. 100 illustrations. xxi + 252pp. 5⅜ x 8. Paperbound $1.50

ATOMS AND MOLECULES SIMPLY EXPLAINED,
B. C. Saunders and R. E. D. Clark
Introduction to chemical phenomena and their applications: cohesion, particles, crystals, tailoring big molecules, chemist as architect, with applications in radioactivity, color photography, synthetics, biochemistry, polymers, and many other important areas. Non technical. 95 figures. x + 299pp. 5⅜ x 8½. Paperbound $1.50

THE PRINCIPLES OF ELECTROCHEMISTRY,
D. A. MacInnes

Basic equations for almost every subfield of electrochemistry from first principles, referring at all times to the soundest and most recent theories and results; unusually useful as text or as reference. Covers coulometers and Faraday's Law, electrolytic conductance, the Debye-Hueckel method for the theoretical calculation of activity coefficients, concentration cells, standard electrode potentials, thermodynamic ionization constants, pH, potentiometric titrations, irreversible phenomena. Planck's equation, and much more. 2 indices. Appendix. 585-item bibliography. 137 figures. 94 tables. ii + 478pp. 5⅝ x 8⅜.
Paperbound $2.75

MATHEMATICS OF MODERN ENGINEERING,
E. G. Keller and R. E. Doherty

Written for the Advanced Course in Engineering of the General Electric Corporation, deals with the engineering use of determinants, tensors, the Heaviside operational calculus, dyadics, the calculus of variations, etc. Presents underlying principles fully, but emphasis is on the perennial engineering attack of set-up and solve. Indexes. Over 185 figures and tables. Hundreds of exercises, problems, and worked-out examples. References. Two volume set. Total of xxxiii + 623pp. 5⅜ x 8. Two volume set, paperbound $3.70

AERODYNAMIC THEORY: A GENERAL REVIEW OF PROGRESS,
William F. Durand, editor-in-chief

A monumental joint effort by the world's leading authorities prepared under a grant of the Guggenheim Fund for the Promotion of Aeronautics. Never equalled for breadth, depth, reliability. Contains discussions of special mathematical topics not usually taught in the engineering or technical courses. Also: an extended two-part treatise on Fluid Mechanics, discussions of aerodynamics of perfect fluids, analyses of experiments with wind tunnels, applied airfoil theory, the nonlifting system of the airplane, the air propeller, hydrodynamics of boats and floats, the aerodynamics of cooling, etc. Contributing experts include Munk, Giacomelli, Prandtl, Toussaint, Von Karman, Klemperer, among others. Unabridged republication. 6 volumes. Total of 1,012 figures, 12 plates, 2,186pp. Bibliographies. Notes. Indices. 5⅜ x 8½.
Six volume set, paperbound $13.50

FUNDAMENTALS OF HYDRO- AND AEROMECHANICS,
L. Prandtl and O. G. Tietjens

The well-known standard work based upon Prandtl's lectures at Goettingen. Wherever possible hydrodynamics theory is referred to practical considerations in hydraulics, with the view of unifying theory and experience. Presentation is extremely clear and though primarily physical, mathematical proofs are rigorous and use vector analysis to a considerable extent. An Engineering Society Monograph, 1934. 186 figures. Index. xvi + 270pp. 5⅜ x 8.
Paperbound $2.00

APPLIED HYDRO- AND AEROMECHANICS,
L. Prandtl and O. G. Tietjens

Presents for the most part methods which will be valuable to engineers. Covers flow in pipes, boundary layers, airfoil theory, entry conditions, turbulent flow in pipes, and the boundary layer, determining drag from measurements of pressure and velocity, etc. Unabridged, unaltered. An Engineering Society Monograph. 1934. Index. 226 figures, 28 photographic plates illustrating flow patterns. xvi + 311pp. 5⅜ x 8. Paperbound $2.00